Managing Small-scale Fisheries

Alternative Directions and Methods

Fikret Berkes, Robin Mahon, Patrick McConney, Richard Pollnac, and Robert Pomeroy

INTERNATIONAL DEVELOPMENT RESEARCH CENTRE
Ottawa • Cairo • Dakar • Johannesburg • Montevideo • Nairobi • New Delhi • Singapore

Published by the International Development Research Centre
PO Box 8500, Ottawa, ON, Canada K1G 3H9

Legal deposit: 2nd quarter 2001
National Library of Canada
ISBN 0-88936-943-7

The catalogue of IDRC Books and this publication may be consulted online at http://www.idrc.ca/booktique.

Contents

Foreword

Human dependence on marine and coastal resources is increasing. Today, small-scale fisheries employ 50 of the world's 51 million fishers, practically all of whom are from developing countries. And together, they produce more than half of the world's annual marine fish catch of 98 million tonnes, supplying most of the fish consumed in the developing world.

At the same time, increased fishery overexploitation and habitat degradation are threatening the Earth's coastal and marine resources. Most small-scale fisheries have not been well managed, if they have been managed at all. Existing approaches have failed to constrain fishing capacity or to manage conflict. They have not kept pace with technology or with the driving forces of economics, population growth, demand for food, and poverty. Worldwide, the management and governance of small-scale fisheries is in urgent need of reform.

One of the reasons for this neglect is that fishery science has largely been devoted to stock assessment, with a geographical focus on countries of the North and a disciplinary focus on biology and, to some extent, economics. As would be expected, such research has not served the fishery-management needs of the South, including countries that primarily depend on small stocks. Also, it has addressed neither the socioeconomic needs of fishing populations nor the potential benefits of more collaborative forms of governance.

In response, Canada's International Development Research Centre (IDRC) has funded research on the theory and practice of small-scale fisheries management in the developing world. Practically as one, this research has shown that to improve the state of fisheries (and, indeed, of most natural resources), managers need to exert more control over access. It has pioneered an interdisciplinary natural and social science of fishery management for the South, including approaches that are driven by management objectives, versions of local framework analysis, and survey methodologies in the mould of participatory rural appraisal. Recommendations have included new governance regimes, such as community-based management or comanagement, and increased use of local fishery knowledge.

This book presents alternative concepts, tools, methods, and conservation strategies, many of which were developed with IDRC support. It shows how to use these methods in a practical way and places a strong emphasis on ecosystem management and participatory decision-making. Natural resource managers, particularly of fisheries and aquatic resources, in developing countries, will find this book very useful, as will managers in other sectors because of the increasing spillover of management approaches across resource sectors. This book will also be of use to representatives from all government agencies, development institutions, nongovernmental organizations, international executing agencies, and donor agencies that are involved in fisheries management, particularly for the small-scale sector.

IDRC's hope is that this book will, in its own way, assist developing-country fisherfolk in their pursuit of a sustainable livelihood. We also hope that it presents a convincing case for a more people-centred model of natural resource management.

Brian Davy
Team Leader, Biodiversity
International Development Research Centre

Preface

Each of the five of us has been working on various aspects of small-scale fisheries over the years. But the story of this particular book really started in 1997 with Robin's paper in the *Canadian Journal of Fisheries and Aquatic Sciences*. In it, Robin pointed out that the conventional approaches to fishery management had been inappropriate for small states and small stocks, and identified the need for a different kind of fishery science better suited to small-scale fisheries management. Fikret and Robin sat under a tropical tree to explore the need for a book-length treatment of the subject, but first, a team had to be assembled and some funding found.

The second step came in May 1998 at the conference of the International Association for the Study of Common Property (IASCP) in Vancouver. The IASCP conference provided the setting for Bob, Patrick and Fikret to get together with IDRC's Brian Davy to discuss the idea. The IDRC came up with funding for an initial meeting to develop the book proposal. In January 1999, Richard joined the group and the whole team met for the first time.

At its January 1999 meeting, the group quickly agreed on the basics. Most of the world's fishery science effort had been devoted to stock assessment, with geographic focus on countries of the North and disciplinary focus on biology and, to some extent, economics. Such fishery science had not served well the fishery management needs of the South, including countries that primarily depended on small stocks. As well, it had not adequately addressed the socioeconomic needs of fisherfolk, nor the potential benefits of participatory management. However, a number of promising approaches had recently been developed and were now available for fishery managers to use.

These included methodological approaches that emphasized management objectives and processes rather than just stock assessment. They included ways of accessing fishers' knowledge to enrich the information available for management, means to build capacity and institutions, and collaborative approaches to include resource users in the management process. The proposed book would be a guide to these alternative management approaches, providing a vision of an ecologically, socially, and economically sustainable small-scale fishery in which management was participatory and the people who did the fishing were no longer politically and economically marginalized.

IDRC accepted the proposal, and work on the book soon began. The authors drafted chapters during the summer and fall of 1999, and met to discuss them in Kingston, Rhode Island, in December 1999. At this meeting, we realized that we were not discussing merely small states and small stocks. These fishery management alternatives were as relevant to the North as to the South, and to medium-scale fisheries as well as to the small. Our focus remained, however, on small-scale fisheries. The Kingston meeting was followed by joint writing sessions involving three of the team members in March 2000 in Barbados. By this time, the original plan that each team member would write one section of the book had been abandoned in favour of a series of chapters, to be written collaboratively, with each member taking the lead for a part of the book. This would make the best use of our interdisciplinary team (fisheries biology, ecology, human ecology, resource economics, and anthropology) and the synergism of having academics and practitioners working together.

We revised our chapters, then distributed and discussed them in a meeting at the World Resources Institute, Washington DC, in June 2000. Each chapter was thoroughly edited (now for the third time), missing pieces were inserted and the gaps filled. The interdisciplinary mix in the team had proved its worth. Our range of expertise with the fishery literature, along with our geographic range of experience with small-scale fisheries worldwide, brought together a rich mix of material for the book.

We thank our institutions for their support during the book project: the Fisheries Division, Ministry of Agriculture of the Government of Barbados; the University of Rhode Island; the University of Manitoba; and the World Resources Institute. The Fisheries Division provided crucially important early feedback on the project concept through a workshop in January 1999. Each one of us have many fishers and communities to thank — too many to mention here. In particular, however, Bob and Richard would like to thank ICLARM and the thousands of fishers in ICLARM projects for the small-scale fishery management experience that they provided. Fikret thanks the Social Sciences and Humanities Research Council of Canada (SSHRC) and the Ford Foundation for making it possible to work with a variety of fishing communities over the years.

For ideas, critique, and assistance, special thanks are due to Bisessar Chakalall, Tony Charles, Janice Cumberbatch, Mina Kislalioglu Berkes, Bob Johannes, and Hazel Oxenford. We pay special tribute to Anton Atapattu of Sri Lanka Department of Fisheries, who passed on prematurely during the project. Anton was an inspiring fishery manager committed to participatory management; he would have enjoyed this book. Jem Berkes provided computer skills for manuscript preparation. Finally, Brian Davy of the IDRC deserves much appreciation for encouraging us and for making the book possible. Thanks are due to IDRC for its support, and particularly to IDRC's managing editor, Bill Carman, and the book production team.

Fikret Berkes, Robin Mahon, Patrick McConney, Richard B. Pollnac,
and Robert S. Pomeroy
(in alphabetical order)

January 2001

Chapter 1
Introduction

1.1 Not just another fisheries book

Although definitions of small-scale fisheries and fisheries management vary widely, it is generally accepted that their goal is to produce for generations of humans a reasonably steady, sustainable stream of benefits from living aquatic resources. However, a glance through current fisheries literature reveals a perplexing array of perspectives and prescriptions to achieve this goal. There are few simple solutions for the problems that fisheries science and management address anywhere in the world. This is particularly so for small-scale fisheries, which this book is primarily about. However, fresh perspectives, methods, and approaches offer new opportunities for small-scale fisheries management. This book is about these alternatives.

1.2 A personal perspective

Through their combined experience, the authors have had personal association with small-scale fisheries in many parts of the world and in various capacities. As fisheries management practitioners, teachers and consultants, we realized that no comprehensive text took the perspective of the small-scale fishery — hence the need for this book. One of us is a full-time manager of small-scale fisheries in the Caribbean. This section is written mainly from his perspective, illustrating how this book addresses the needs of a fairly typical small-scale fisheries manager. This profile highlights some of the points and principles that are relevant to most, if not all, small-scale fisheries. Managers elsewhere may not have experienced everything from this perspective, but most will find points that apply to their situation, as will many managers of larger-scale fisheries. These points are the common ground upon which to build shared interest in alternative directions for small-scale fisheries management.

1.2.1 Vision of a small-scale fishery

Formalized fisheries management does not have a long history in the eastern Caribbean (Mahon 1990), and it was only recently introduced to Barbados (McConney and Mahon 1998). Eight fisheries, most multispecies and a few essentially single species, were identified for management. None had the benefit of detailed stock assessment within national jurisdiction prior to the introduction of management. Assessing fish stocks was judged neither feasible nor necessary. We adopted a precautionary approach.

One of the first steps in the management process was to agree through multistakeholder processes on a shared vision of the fisheries, setting out a clear picture of what the fishing industry of the future would look like if the fishery was successfully managed. In Barbados, we opted for separate but integrated visions of the harvest sector, postharvest sector, and the state so that most of the critical stakeholders could have a vision with which to identify (Fisheries Division 1997). Small-scale fisheries stakeholders and managers elsewhere may consider this Barbados vision worth attaining:

Small-scale fisheries not marginalized either by larger scale fisheries or other sectors of the economy, but endowed with the human, institutional, physical, and financial resources necessary for proper management. Participatory, empowering management, with diverse stakeholders reaching consensus on the objectives that drive management, and on the means to resolve conflicts. Responsible fisheries ensuring ecosystem and human system sustainability or rehabilitation under conditions of uncertainty in order to maintain or improve quality of life for generations to come.

Having created and agreed upon the vision, the next steps were to identify the issues to be addressed and objectives to be achieved in order to get there. This included alternative strategies and implementation plans using the most appropriate fisheries management approaches and tools. Here the process got difficult.

1.2.2 How to get there is not clear

For a small-scale fisheries manager, the path toward the vision can be strewn with a surprising array of obstacles and pitfalls. The first, in this case, was politics. People in Barbados were accustomed to open access to fisheries resources, with little management or conservation intervention from the state despite token fisheries regulations. Policy and practice had been oriented toward increased fish exploitation, not conservation. Policymakers thought that increased fishing effort would produce increased harvests, ignoring evidence to the contrary in many parts of the world. Many people in the fishing industry considered fisheries to be inherently unpredictable and unmanageable. Decisions on fisheries matters were made with little reference to resource management except where overfishing was blatantly obvious. The politically appropriate, rather than the scientifically appropriate, directed decision-making in Barbados, as elsewhere in developed countries (Ludwig *et al.* 1993).

But this was not just a case of science and management being intentionally ignored. Most people who fish, some decision-makers, and the general public lacked knowledge of fishery science and management (McConney 1997). Without appreciation of the benefits that these analytical perspectives could bring, people believed that common sense alone was an adequate basis for fisheries decisions. It was necessary to convince both those who fish and those who make policy that fisheries science and management were worthwhile. This also meant that it would be worth their while to participate in the consultative fisheries management planning process designed as the opportunity for all stakeholders to exchange information in order to achieve consensus and make decisions. Barbados, like many other Caribbean fisheries authorities, wants to share power and responsibility with fisheries stakeholders in comanagement as a means of survival within the governance system (Chakalall *et al.* 1998).

Without the willing collaboration of the fishing industry, small-scale fisheries managers stand little chance of carrying out their mandates. This dependence of small-scale fisheries management units on their clients emphasizes their need to effectively communicate, plan, and work with the fishing industry rather than continue the present command-and-control approach. A prerequisite to working efficiently with the fishing industry was to deal with fishers' organizations, such as associations and cooperatives, at the community level. A program was undertaken to develop organizations that could play a meaningful role in fisheries management (McConney in press). Thus, both the managers and the managed were empowered.

Whether the management approach was participatory or not, the question of appropriate fisheries science and management tools would still have arisen. In Barbados, like many small island developing states, the capacity for state management is severely constrained, with financial, human, and physical resources all being scarce. In this region, fisheries scientists and managers (often, the same person is both) are acutely aware of these constraints, since most have been trained in North America or Europe. They know that, for a number of reasons, much of the conventional fishery science they have been taught is not feasible in their situation. Governments in most developing countries are not likely to expend scarce resources on the conventional means of researching and managing small-scale fisheries. This is because small-scale fishing is usually seen as a social safety net, cultural feature, and source of employment for the less skilled or educated, not as a major engine of the economy.

In the case of Barbados, eight fisheries were identified for management. Most were multispecies and multi-fleet, seasonal, and shared-resource small-scale fisheries. Does practical guidance on how to manage such fisheries exist? Our Fisheries Division cannot afford subscriptions to mainstream fisheries science and management journals. And even if it could, Barbados-type fisheries are poorly represented in the mainstream literature, which focuses on large-scale fisheries or temperate small-scale fisheries (Pauly 1994). Although some journals and grey literature reports address tropical small-scale fisheries at no or low cost to the subscriber, few publications adequately cover the methods of small-scale fisheries management appropriate for everyday use by small fisheries authorities. And while useful fisheries information is increasingly available via the Internet, piecing together what is required to guide small-scale fisheries management continues to be a challenge.

Barbados had one fisheries biologist, with little field and technical support, to deal with all of the scientific aspects of the eight fisheries, using fishery-dependent data almost exclusively. Analytical methods that were sufficiently simple and robust to handle considerable variation in data quality, given the capability constraints, had to be sought. At the beginning, collection of social and economic data was minimal because fisheries management was equated with use of stock assessment outputs, which were non-existent. Now, conventional fisheries science and management have methods to deal with poor data and other deficiencies, using an ordinary microcomputer. However, these methods are still relatively data-, time-, finance-, and, computationally intense — a challenge to Barbados' institutional capability.

It was not possible to give adequate attention to all of the fisheries all of the time using conventional approaches. Furthermore, such expenditure of effort on assessment and provision of information had to be evaluated in the context of the likely basis for decision-making and the value of the fisheries. These factors pointed to the need for alternatives to conventional approaches. In conventional approaches, fisheries science, management and development tend to be separate from each other, and also separate from the wider body of knowledge held by the fishing industry and stakeholders with linkages to other sectors of the economy. The alternative approach combines all parts of the fisheries system.

Barbados used a new, people- and objectives-oriented approach, combining common sense and ecological knowledge about fisheries with the specialized analytical knowledge of fisheries science and management (McConney 1998). This approach, which is still evolving, has these characteristics:

- Management benefits the common good of the public rather than partisan interests.
- Common sense guides practical action, which is driven by objectives.
- The approach is based on situation-appropriate fishery science, using mainly simple methods.
- Analytical principles of science (social and natural) are applied to problem solving.
- Scientific knowledge is a complement to, not a substitute for, traditional knowledge. Common ecological knowledge, including that of people who fish, is crucial.
- Common property issues in fisheries management are addressed.
- Stakeholders' common interests are the basis for negotiation or consensus.
- The management system is accessible to average people so that they can participate in a meaningful way. This includes the use of common language rather than scientific jargon.

None of the above is new or revolutionary. But together, these principles create a powerful and as yet underutilized approach to fisheries management. This is only a sample of the many alternative directions now available for small-scale fisheries management, accompanied by a suite of rigorous methods appropriate for small-scale fisheries science.

1.2.3 The need for new and alternative directions

Most of the world's fishery science has been devoted to stock assessment, with geographic focus on countries of the North. The disciplinary focus has been on biology and, to some extent, economics. Without the inclusion of much social science, conventional approaches have not adequately addressed the socioeconomic needs of fishing populations and the potential benefits of collaborative governance. People were at the periphery, not the centre, of conventional fisheries management. Such fishery science has not served well the fishery management needs of the South, including countries that primarily depend on small stocks, often exploited by small-scale fisheries on a community basis. The Caribbean situation described above is not at all unique. Authors have referred to crises in fisheries (for example, McGoodwin 1990; Buckworth 1998) and the consequent need to reinvent fisheries management (Pitcher *et al.* 1998). This is largely due to the inadequacies and failures of conventional fishery management as applied to both developed and developing countries.

However, a number of promising new or revised approaches are now available to fishery managers in developing countries. These include methodological approaches that emphasize fishery management objectives and participatory decision processes rather than the customary primary focus on fish stock assessment and population dynamics with a secondary focus on the human dimensions of the fishery. Included are new governance regimes, such as community-based management and comanagement that have the potential to address community development as an integral part of fishery resource management. Interdisciplinary and social science methodologies feature prominently. These include versions of logical framework analysis, the use of fishers' knowledge of local ecology, and participatory rural appraisal types of survey. Integrated coastal area management may incorporate fisheries issues into the total scheme of coastal economic development using a geographic information system (GIS), providing a powerful visual tool for decision-making and conflict resolution.

No-cost or low-cost journals and grey-literature reports that address tropical small-scale fisheries provide invaluable information to the small-scale fisheries manager. Cost-effective use of Internet resources can, with relative ease, put fishery managers in remote locales in touch with global data resources, personal contacts, information exchanges, funding opportunities, and more. Computers and user-friendly software allow even the smallest fisheries authority to present information to decision-makers in a simple but comprehensive manner using attractive text and visual images. These new alternatives, added to the existing fisheries management toolbox, furnish the small-scale fishery manager with methods appropriate to data-poor, human- and financial-resource-limited situations. Information on these alternatives and other emerging approaches is scattered throughout the literature but appears only rarely in mainstream texts and never as an integrated package.

1.3 Scope of the Book

This book is a synthesis of practical and appropriate approaches to fisheries science and management of small-scale fisheries. Many of these can be called alternative approaches or new directions because fisheries managers rarely use them now. The information is presented in a practical, readable format to serve practising fishery managers, fishery-related government and non-governmental organizations, and institutions that provide graduate-level training to people involved in fisheries management. Illustrations and real-life examples appear throughout the text.

Many small-scale fisheries managers have been trained in or influenced by conventional approaches to fisheries management — methods that tend to be data intense, complex and costly. Methodology often poses significant challenges and constrains management initiatives. That is why this book pays special attention to the application of methods.

This book focuses on small-scale fisheries in developing countries: the people who fish, their communities, and the linkages within and outside the fisheries systems. Some of this information is also relevant to small-scale fisheries in developed countries and to large-scale fisheries everywhere. The content addresses capture fisheries (freshwater, floodplain, estuarine or marine) but not aquaculture. The book is about fisheries management rather than administration and development (but the authors acknowledge that this distinction is somewhat artificial to the practising fisheries manager).

Next, this chapter briefly looks at small-scale and other types of fisheries, then sketches the history of current fisheries science and management.

1.4 Types of fisheries

Those preparing to manage small-scale fisheries must first be aware of the diversity of types and size of fisheries. This subject alone could fill a book. Since such information is widely available in the literature, this book merely summarizes that diversity.

What is a fishery? Different categories and descriptions exist for various purposes. Pitcher *et al.* (1998) define a fishery only by resource and gear type — a minimal picture. A more comprehensive description might include all the categories of fishers, along with the types of gear they use, exploiting a particular resource. This would correspond to the harvest sector entity requiring assessment and management as a whole: the fisheries management unit. The postharvest sector, which includes buyers, processors, and market linkages, is also part of the picture, as is the fishery governance system of state and civil society. Thus, a fishery has biological, technological, economic, social, cultural, and political dimensions.

Various terminologies are used to label the range of fisheries (**Table 1.1**). The terms differ in the details of definition but not in substance. It is useful, however, to distinguish the large-scale (commercial/industrial) from the small-scale (commercial, artisanal, subsistence) ends of the spectrum. Strictly speaking, all fisheries are commercial. Even the smallest artisanal fishery sells what is surplus to household needs. Today there are very few fisheries in which none of the catch is sold, and these are usually termed subsistence fisheries. In such fisheries, cash transactions are minimal, but fish tend to be traded or shared extensively among kinship and social networks. These, too, are part of small-scale fisheries.

Table 1.1. Categories and dimensions of fisheries.

Fisheries-related characteristics	Categories		
	Large-scale	Small-scale	Subsistence
	Industrial	Artisanal	
Fishing unit	Stable, with division of labour and career prospect	Stable, small, specialized with some division of labour	Lone operators, or family or community group
Ownership	Concentrated in few hands, often non-operators	Usually owned by senior operator, or operators jointly, absentee owner	Owner-operated
Time commitment	Usually full-time	Either full-time or part-time	Most often part-time
Boat	Powered, much equipment	Small; inboard motor (or small outboard)	None, or small, usually non-motorized
Equipment types	Machine-made, assembled by others	Partly or wholly machine-made materials, often operator-assembled	Often hand-made materials, operator-assembled
Gear sophistication	Electronics, automation	Mechanized and manual	Mainly non-mechanized
Investment	High; large proportion other than by operator	Medium to low; entirely by operator	Low
Catches (per fishing unit)	Large	Medium to low	Low to very low
Disposal of catch	Sale to organized markets	Organized local sale, significant consumption by operators	Primarily consumed by operator, his family, and friends; exchange by barter; occasional sale
Processing of catch	Much for fishmeal and non-human consumption	Some drying, smoking, salting; primarily human consumption	Little or none; all for human consumption
Operator's income level	Often high	Middle to lowest brackets	Minimal
Integration into economy	Formal; fully integrated	Partially integrated	Informal; not integrated
Occupationality	Full-time or seasonal	Often multi-occupational	Multi-occupational
Extent of marketing	Products found worldwide	Often national and local	Local or district-level only
Management capacity of fisheries authority	Considerable, with many scientists and managers	Minimal to moderate, with few scientists/managers	Often not managed except by the resource users
Management units	One or few large units	Usually many small units	Very many small units
Fisheries data collection (also see ***Figure 1.1****)*	Not too difficult, given the authority's capacity	Difficult due to fisheries and authority's features	Often no data may be collected due to difficulty

Source: adapted from Smith 1979

1.4.1 Large-scale commercial fisheries

Large-scale commercial fisheries (also referred to as industrial fisheries) land a large proportion of the world's fish catch from a relatively small number of fish stocks or subpopulations. These fisheries are highly mechanized, using large, technologically sophisticated vessels and equipment, often with on-board processing. Large-scale commercial fisheries generally exploit large stocks of widely distributed species in productive areas. These fisheries tend to target the following groups of species:

- Wide-ranging, oceanic, large pelagic species, using surface longlines, purse seines, and so on;
- Demersal fishes of highly productive shelves and slopes, using trawls and bottom longlines;
- Schooling small pelagics, such as clupeoids and mackerels of highly productive upwelling and river outflow–affected systems, using purse seines or pelagic trawls;
- Shrimps of tropical river outflow–affected shelves, using trawls.

These large fisheries are the ones for which the most data are available, and are therefore the best understood. Management tools and processes, developed mainly with these fisheries in mind, are characterized by established operational management systems, even if these have often failed to prevent overexploitation. Developed countries employing conventional fisheries science and management predominantly pursue these types of fisheries. However, developing countries may also undertake them through joint ventures that are often heavily weighted in favour of the developed-country partner.

1.4.2 Small-scale fisheries

No universal definition of small-scale fisheries exists, and other terms such as traditional or artisanal are sometimes used synonymously. In trying to define "small-scale," several authors concluded that it can differ according to location and context (Smith 1979; Panayotou 1982; Berkes and Kislalioglu 1989; Poggie and Pollnac 1991a) but there are features in common to most small-scale fisheries. This comparison of small and large scales of fisheries highlights the global importance of the small-scale fisheries. **Table 1.2** illustrates this comparison.

Table 1.2 Large-scale and small-scale fisheries compared.

Key features of the fisheries	Large-scale fisheries	Small-scale fisheries
Direct employment in fishing	500 000 people	50 000 000 people
Fishery-related occupations	—	150 000 000 people
Fishing household dependents	—	250 000 000 people
Capital cost per fishing job	US$30 000 – $300 000	US$20 – $300
Annual catch for food	15 – 40 million tonnes	20 – 30 million tonnes
Annual fish bycatch	5 – 20 million tonnes	< 1 million tonnes
Annual fuel oil consumption	14 – 19 million tonnes	1 – 2.5 million tonnes
Catch per metric tonnes of oil used	2 – 5 metric tonnes	10 – 20 tonnes

Small-scale commercial fisheries exploit many of the same stocks as are exploited by the large-scale commercial fisheries but also exploit a large number of smaller stocks. They may be highly modernized and technologically sophisticated. Such fisheries tend to target the following groups of species:

- Deep demersal fishes of tropical shelf slopes, typically using nets, lines and traps;
- Coastal large pelagic fishes, typically by trolling or with small-scale longlines;
- Coastal demersal fishes of temperate shelves and bays, using traps, nets, and longlines, often exploiting the same stocks as large-scale trawl fisheries operating further offshore but frequently targeting different life-history stages.

Traditional, artisanal, and subsistence fisheries are also in the category of small-scale fisheries, exploiting many of the stocks harvested by commercial fisheries. In addition, they exploit a great variety of very small stocks distributed over numerous management units (**Figure 1.1**). Some of these fisheries are mechanized but most use traditional fishing gear, such as small nets, traps, lines, spears, and hand-collection methods. Of all the fisheries, biodiversity of the catch is highest in these. For that reason, and because low gear used is unselective, these harvests include a greater variety of species than do those of the larger commercial fisheries. Traditional, artisanal, and subsistence fisheries tend to target the following groups of species:

- Fishes and invertebrates of coral reefs, typically with traps, spears, lines, and by hand;
- Fishes and invertebrates of coastal lagoons and estuaries, typically using nets;
- Stream and river fisheries, typically using nets;
- Aquarium species in all habitats, using nets and noxious substances.

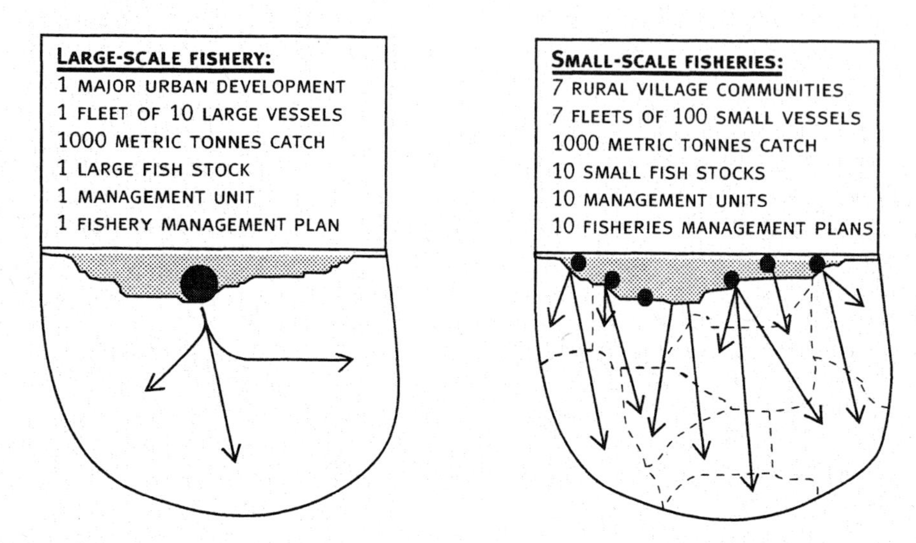

Figure 1.1 Relative complexity of large-scale and small-scale fisheries.

Small-scale fisheries tend to predominate in tropical, less-developed areas, where fisheries (and environmental) management capacity may be poorly developed or even non-existent. However, they are also common in coastal areas of developed countries, such as along the Atlantic coast of Canada and the USA. The smallest stocks seldom have a total value (as distinct from unit value) that is high enough to support conventional information collection and management systems. Many small-scale commercial fisheries are well documented when they coincide geographically with stocks that support large commercial fisheries and thus receive attention from the scientists and surveys focusing on the larger stocks. Others, though, are in developing areas of the world, or in remote parts of the developed world, such as northern Canada or Siberia, where little fishery assessment is done and management systems are weak or non-existent.

Artisanal and subsistence fisheries are seldom well documented, even when they occur side-by-side with large-scale commercial fisheries. Though little studied, they are the predominant fishery in tropical developing countries (King 2000). Subsistence fisheries may contribute 80 percent of the catch in some Pacific islands (Dalzell *et al.* 1996). The targeted stocks also tend to be the smallest in abundance and the least widely distributed, particularly when those in fresh water are considered. With a few exceptions, such as seahorses and some aquarium trade species, the species of these fisheries do not attract much attention from fisheries scientists or conservationists and thus have a low public profile.

The small-scale fishing industry has other characteristics that managers need to take into account. In both developing and developed countries, the number of small-scale vessels and fishermen exceeds those in the large-scale industry. This multitude of small enterprises leads to issues of collective action, power, and conflict resolution.

In addition, small-scale fisheries are often based in small coastal communities that depend on local resources that can be affected, positively or negatively, by surrounding economic activities. These populations may be geographically remote in large countries, and can be politically distant from the centres of fisheries decision-making, even in small islands. Small island developing states with shared fisheries resources are prevalent in the Caribbean and Pacific regions. The multispecies and multi-fleet harvest sectors common in tropical areas make the task of managing small-scale fisheries more challenging there than in the ecologically less complex north. It has been noted that the fisheries authorities in developing countries may be limited in their capacity to manage small-scale fisheries. This incapacity may be caused by their use of conventional fisheries management methods that were developed in the north and do not suit small-scale fisheries.

1.5 Review of fisheries management from a "people" perspective

1.5.1 Is fisheries management necessary?

It is generally accepted that without management, the benefits that most fisheries produce will diminish. This is the "tragedy of the commons" (Hardin 1968) argument, and it is now clear that a tragedy *will* occur in the absence of management, whether that management come from central government or local communities. In many cases, the resources will even become commercially extinct (that is, even though some members of the species survive, they are not worth fishing for). In extreme cases, they may become biologically extinct (Roberts and Hawkins 1999). This possibility of biological or economic extinction has only recently been appreciated. Before the turn of the 20th century, the industrialized countries of Europe believed that fishery resources were inexhaustible.

Current literature is full of examples supporting the warning that unmanaged fisheries will lose their economical viability or even collapse. Most relate to the dramatic fall-off of large fish stocks. The Peruvian anchoveta, northern cod, New England groundfish, bluefin tuna and Atlantic swordfish are some of these (Buckworth 1998). However, numerous small stocks have suffered similar fates without attracting much attention. Some examples of these are reef fish stocks in many tropical countries and queen conch in the Caribbean (FAO 1993).

The spectre of extinction is much less documented. Although only a few clear-cut cases of marine extinction can be linked to exploitation (Roberts and Hawkins 1999), concern about this possibility can be expected to increase. The threat of extinction is probably higher in the tropics than in temperate areas because the tropics are home to more species with smaller population sizes. Information on the role of fisheries in extinction in tropical habitats is scarce, but traditional local management may have helped maintain some small-scale fisheries; for example, by the use of reef and lagoon tenure systems.

In the Pacific, for example, realization that fisheries resources are not inexhaustible has led, after centuries of exploitation, to the development of social and cultural systems of fisheries management (Johannes 1978). Scientists and managers are now aware of the ubiquity, effectiveness, and efficiency of traditional systems and are beginning to use similar concepts in modern management. The techniques used in the two approaches tend to be remarkably similar, except that customary methods are embedded to a much greater extent in social systems than are modern ones (Ruddle 1988). Although use, not necessarily conservation, was the objective of traditional systems, sustainable resource management was the effect. Such traditional systems had adaptive value because they helped to maintain the benefits to users over generations, sometimes as a matter of social, economic, and biological survival (Berkes 1999). Principles of sustainable use and precautionary approaches, now embodied in the 1982 *United Nations Convention on the Law of the Sea* and related fisheries agreements, are not new.

From this global perspective, there appears to be consensus on the necessity of fisheries management.

1.5.2 Management approaches

The goals of management are, first, to prevent biological and commercial extinction, and, second, to optimize the benefits derived from the fishery over an indefinite period; in summary — the goal is to use resources sustainably. This goal encompasses a great deal of complexity. Assessing the risk of biological extinction is the focus of ongoing debate in the international natural resources management arena (for example, The World Conservation Union [IUCN], CITES, and the Food and Agriculture Organisation [FAO]). Fisheries management has focused for decades on avoiding commercial extinction and optimizing benefits.

Most of the fishery science themes and concepts that influence fisheries managers are associated with modern, conventional approaches. It is instructive to observe how these approaches' management objectives have changed over time — such objectives as maximum sustainable yield (MSY) (Larkin 1977), maximum economic yield (MEY) and optimum sustainable yield (OSY) (Roedel 1975). These changes were accompanied or instigated by changes in understanding of fisheries systems (and willingness to admit ignorance) and by scientists' and managers' attempts to model nature (Panayotou 1982). Uncertainty and complexity are now acknowledged and addressed in various ways, some of which incorporate the human dimension. It is even fashionable to say that "we should manage people, not fish," but there is little evidence of this cliché becoming the focus of conventional fisheries approaches.

We can review these approaches from many different angles, but the one chosen here examines them from the perspective of how people (harvesters, decision-makers and society) fit in. In order to keep this review brief and on focus, the authors do not explain basic concepts and models in detail. Elaborations are available in some of the references, such as Panayotou (1982), and in the glossary at the end of this book.

1.5.3 What does fisheries management yield?

The output from a fishery is often referred to as its yield. This can be measured in several ways, such as quantity of fish harvested (biological), revenue from the fishery (economic), or a composite and more intangible "benefit to society" (social and cultural). Maximum sustainable yield (MSY) looks at the biological measure of fish harvested, shown in a variant of a typical static bio-economic illustrative diagram (**Figure 1.2**).

MSY is based on information from stock assessment, irrespective of the fisheries model used. Although the illustrative model is static, with computers it is possible to use complex stochastic and dynamic models to derive results that take environmental and other uncertainties into account. The latter make MSY more suitable as a Limit Reference Point (LRP) than a Target Reference Point (TRP) or management objective. This is because overshooting MSY puts the fishery in trouble, while underachieving provides a margin of safety (Caddy and Mahon 1995). These matters are dealt with later in detail, so are not expanded on here.

Fish, not people, figure most prominently in MSY-type biological approaches. A common failure of these has been to overemphasize the fish, often in single-species models, while ignoring the environment and people. Although more recent ecosystem-based models offer more promise on the ecological front, researchers still do not adequately incorporate human predatory behaviour, including market-driven exploitation, into the ecosystem equations. MSY-dominated approaches are associated with command-and-control input regulations that the harvest sector seeks to circumvent, therefore, raising costs of administration and enforcement to obtain compliance.

Maximum economic yield (MEY), on the other hand, does incorporate assumptions about human behaviour, although not necessarily the appropriate assumptions. MEY is biologically more conservative than MSY (**Figure 1.2**). Economic measures used in managing fisheries include taxes and quotas. Individual transferable quotas (ITQs) are popular today in many developed countries but do not suit most developing countries due to many of the features of small-scale fisheries described earlier in this chapter. MEY seeks to maximize the rent from the fishery and therefore the total economic benefit to society while preventing the "tragedy of the commons" (Hardin 1968). The latter is explained later in this book. But the economic assumption that fishers are unfettered individual profit maximizers leads to the conclusion that all profit from the fishery will be dissipated unless managed, preferably through privatization or sole stewardship by the state. This is a gross oversimplification, even though there is considerable validity to the concern about increased fishing effort eroding both rent and biological viability. There is also agreement that property rights are important in fisheries management. Open access is undesirable but, here again, the exclusion of local-scale institutions has narrowed the fisheries management perspective. To ignore management at the communal level is a serious oversight, as is illustrated by community-based successes that outperform the economic prescriptions.

The obligation to manage fisheries using best available information relates not only to biology and economics but also to the social, cultural, and political components

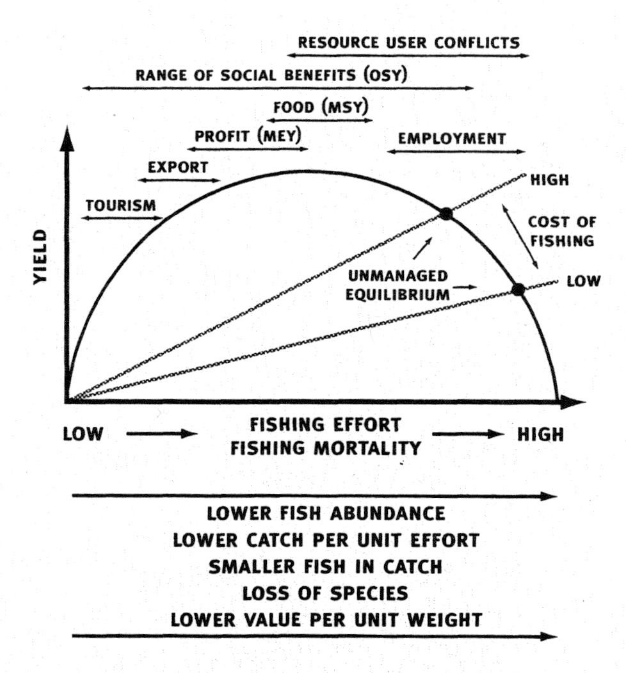

Figure 1.2 Fisheries yields and objectives.

of the fisheries system. Optimum Sustainable Yield (OSY) incorporates the latter components to arrive at yield targets based on management objectives that are broader than the previous two. Examples of different objectives and the areas on the model that they may include are shown in **Figure 1.2**. The idea of optimal yield from a fishery emerged as it became evident that the benefits to be derived from fisheries could be measured in many ways other than simply the weight or the landed value of the catch (Roedel 1975). Consideration of the rather vague concept of optimal sustainable yield was further reinforced when it became clear that maximum sustainable yield as defined by the biological models was, in fact, an unachievable target (Larkin 1977).

The problem is that multiple objectives are messy and OSY rather vague. Maximization of a single objective is much easier than optimization, which, by definition, must address trade-offs and compromises, and these can be difficult. However, the process of reaching consensus on the most appropriate objectives normally brings people into the model far more explicitly than before. Previously, conventional fisheries management and fisheries science held that both the problems and solutions could be clearly specified once sufficient data were plugged into the right stock assessment model. Like a single dart aimed at a distinct target (**Figure 1.3a**), a management measure was supposed to precisely address an equally clear fisheries stock assessment-driven problem. By contrast, a management objective–driven mode

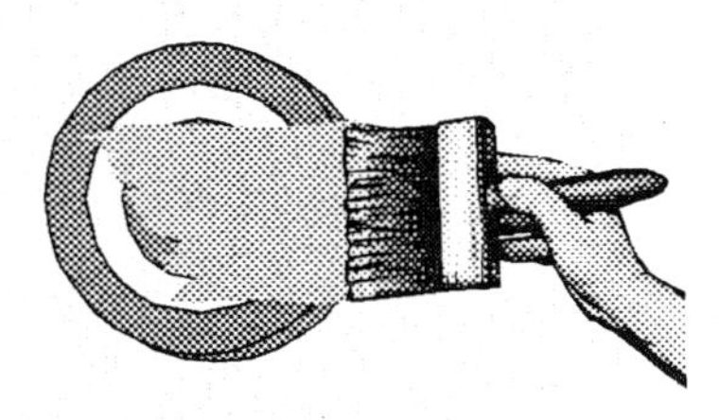

Figure 1.3 Narrow arrow (a) and broad brush (b) approaches to fisheries science and management.

uses a broad-brush perspective of science and management to find creative and innovative solutions to fisheries problems. This paradigm acknowledges that both the questions and answers are plagued with fuzziness, uncertainty, and complexity. Measures that have the breadth of flexibility and adaptability are applied to situations that may themselves cover a spectrum of possible scenarios (**Figure 1.3b**).

It is up to the fisheries governance system, but particularly the fisheries managers, to define what is optimal for a fishery within the boundaries set by sustainability. Recognizing this, more attention is likely to be placed on multi-dimensional indicators for sustainable development that will incorporate information from stakeholders and science (FAO Fishery Resources Division 1999). Much of this book is about the challenge of determining what is optimal and sustainable in a particular set of circumstances. How we approach this will depend to a large extent on our perceptions of the following:

- Who are the managers?
- Who benefits from management?

1.5.4 Who manages for whom?

In most countries, wild fisheries resources are owned by the public, and need to be managed by the state for the benefit of the citizens. The state agency that takes the lead in managing the fishery does so on behalf of a public that may wish to have its say in management decisions. A healthy fishing industry, in which the primary users of the resource (the fishers, traders, and processors) are able to sustain a decent standard of living and return on their investment, is obviously in the best interest of a country. However, the interests of the resource users and of the public do not always coincide, particularly when short-term interests predominate. When this is the case, the government agency leading the management must be prepared to maintain the balance between the interests of users and the public while ensuring that the fishery system as a whole is sustainable. As this book shows, the state can manage a fishery through a variety of arrangements. The authors present and describe several of the alternative approaches to dealing with the problems of small-scale fisheries.

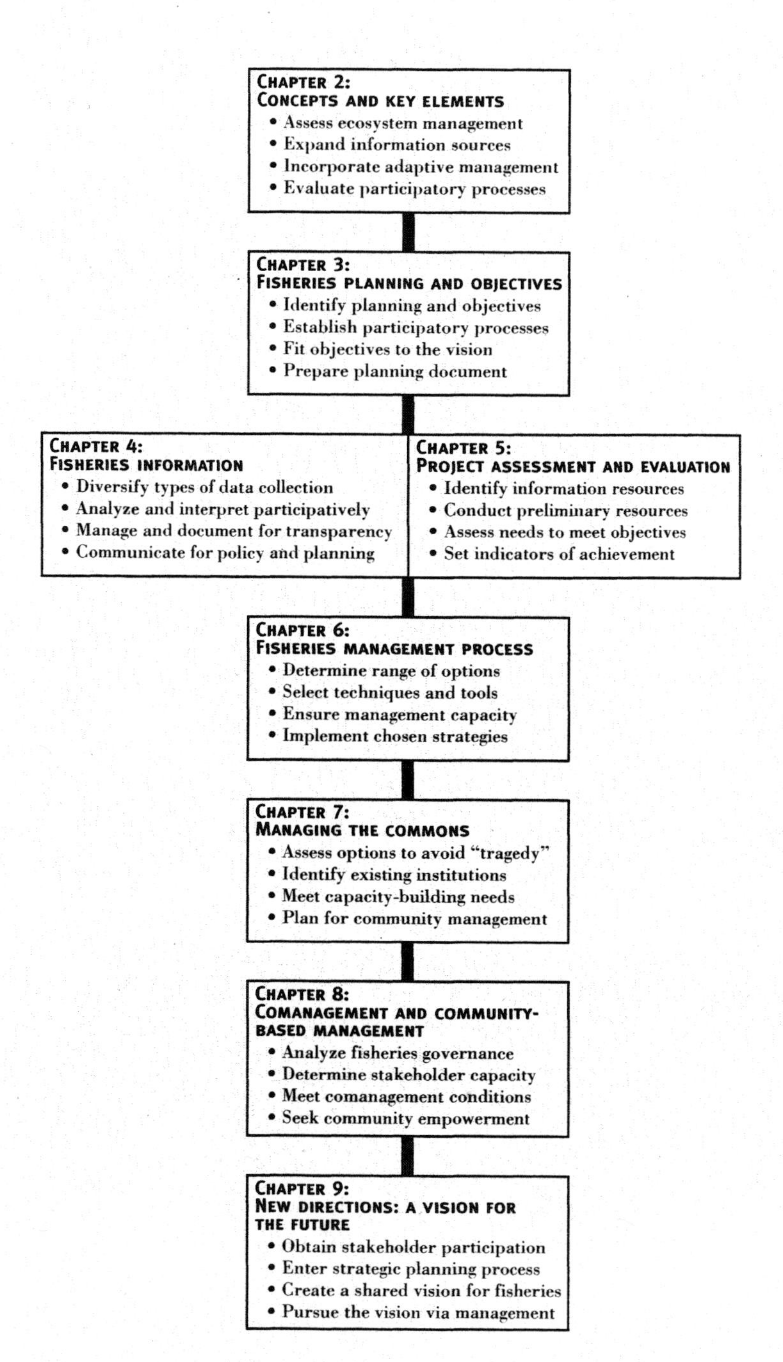

Figure 1.4 Interconnections between chapters.

1.6 What comes next

This section is a schematic introduction to some of the main points addressed in the chapters that follow, and shows how they are interconnected (**Figure 1.4**).

Chapter 2 introduces a number of this book's concepts and perspectives, some well accepted and some controversial. The next four chapters deal with these aspects of fisheries management: planning and objectives (Chapter 3), information (Chapter 4), project assessment and evaluation (Chapter 5), and management process (Chapter 6), with emphasis on new and emerging ideas that can be particularly useful in small-scale fisheries. Chapter 7 highlights management of the commons and the lessons of the commons perspective. Chapter 8 develops this perspective further, focusing on comanagement and community-based management. In the final chapter, the authors share their visions of future directions for small-scale fisheries management.

Chapter 2
Key Concepts in Fisheries Management

2.1 Introduction

Many resource managers are familiar with the view that those who manage marine ecosystems and fisheries management should adopt more holistic approaches. Many also know about the global trends toward management at the local-level and participatory planning. It is not surprising that these two trends have coincided to lead to a new kind of resource management that no longer spends a disproportionate amount of effort on stock assessment, but rather pays attention to some of the other critical dimensions of fishery management. Many of these dimensions are social, especially in the case of small-scale fisheries.

There has been a merging in our thinking of natural systems and social systems. Many discussions of ecosystem-based management now explicitly include humans in the "system" instead of trying to separate them out. In turn, social scientists are broadening their models to include the environment. We use the term social-ecological system to emphasize the point that social systems and ecological systems are in fact linked. The delineation between social and ecological (and between nature and culture) is artificial and arbitrary (Berkes and Folke 1998). The emerging view of an integrated social and ecological system also applies to the world of fisheries (**Figure 2.1**). Globalization issues, such as the vulnerability of local fisheries to international markets, have emphasized the interconnected nature of the world. The manager of small-scale fisheries can no longer ignore environmental movements, biodiversity issues, eco-labeling, and international codes of conduct.

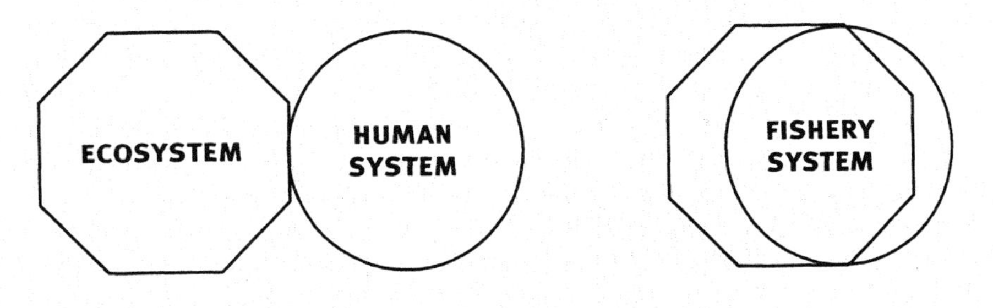

Figure 2.1 Conventional and emerging views of ecosystems and human systems.

Part of this change to bring nature and culture back together again is inspired by the lessons learnt from users of common property resources, such as fisheries, and many of these lessons come from the developing world. We know now, what we did not know until the 1970s and the 1980s, that communities of fisherfolk, in certain cases, are capable of using their resources on a sustainable basis over long periods of time. The literature on common property resources has established that communities of users do not require central government regulations to make and enforce simple and practical systems of resource use. Such findings have helped emphasize that resource managers can deal with users as part of the solution rather than part of the problem. This does not mean that the role of the manager has ended; it means that the role of the manager has changed.

Interdisciplinary social-ecological management of fisheries is not a luxury but a necessity when dealing with complex systems. As an AAAS panel noted, "phenomena whose causes are multiple, diverse and dispersed cannot be understood, let alone managed or controlled, through scientific activity organized on traditional disciplinary lines" (Jasanoff *et al.* 1997). Managing fisheries with an eye to both the biophysical environment and the social and economic environment makes the task of the manager both easier and more difficult. Easier, because such an approach brings the task of management closer to the reality of fisheries and fishermen. More difficult, because such an approach requires a working knowledge of concepts and fields not covered in the conventional education of the resource manager.

The purpose of this chapter is to familiarize the reader with a number of concepts and key elements of the kind of fishery management that this book is about. To replace the "old way of management" with the new, as discussed in Chapter 1, the manager needs to be familiar with the evolving views of marine ecosystems and ecosystem-based management, including uncertainty and risk, the precautionary approach, and marine protected areas. This chapter also deals with how managers apply learning-by-doing, or adaptive management, with possible ways of expanding the sources of fishery management information, including the use of fishers' local or traditional knowledge. Finally, the chapter covers several important topics regarding the social system, such as governance regimes, participatory approaches, comanagement, and empowerment. Many of these key concepts introduced in Chapter 2 will be discussed in more detail in the chapters that follow.

2.2 Ecosystem-based fishery management

Ecosystem-based management can be an important complement to existing fisheries management approaches. As well, several elements of this approach are significant in changing the way we view ecosystems. The *Report of the Ecosystem Principles Advisory Panel of the USA* (EPAP 1999) lists the following as embodying the key considerations and elements of ecosystem-based management of fisheries:

- The ability to predict ecosystem behaviour is limited.
- Ecosystems have real thresholds and limits which, when exceeded, can cause major system restructuring.
- Once thresholds and limits have been exceeded, changes can be irreversible.
- Diversity is important to ecosystem functioning.
- Multiple scales interact within and among ecosystems.
- Components of ecosystems are linked.

- Ecosystem boundaries are open.
- Ecosystems change with time.

The goal of ecosystem-based management, according to the Panel, is to "maintain ecosystem health and sustainability." Both of these concepts warrant some attention. Regarding "ecosystem health," the Panel recommends that indices of ecosystem health can be developed and used as a management target.

> *Ecosystem health refers to a balanced, integrated, adaptive community of organisms having a species composition, diversity and functional organization that has evolved naturally. Provided that a healthy state can be determined or inferred, management should strive to generate and maintain such a state in a given ecosystem. Inherent in this management strategy would be specific goals for the ecosystem, including a description of "unhealthy" states to be avoided (EPAP 1999).*

A related ecosystem prescription is "ecological integrity," usually defined in terms of the maintenance of ecosystem structure and function. Biodiversity is a good measure of "structure," and allows the use of more specific indicators, such as the percentage of long-lived and high-value species, such as groupers, in the catch. "Function" refers to ecosystem processes such as production, energy flow, and nutrient cycling. Many fishery managers use fish yields as an indicator of ecological integrity or ecosystem health: if fish yields decline sharply, this is usually a good indication that something is wrong. Regarding "sustainability," it is important to note that the Panel is referring to ecosystem sustainability and not to single-species yields. As noted by a US National Research Council committee,

> *It is the perception of many observers that single-species fishery management has failed, and that a new approach, which recognizes ecosystem values, is required to achieve sustainable fisheries. A move toward fishing and management that recognizes the importance of species interactions, conserves biodiversity, and permits utilization only when the ecosystem and its productive potential is not damaged, is a worthy objective (NRC 1999).*

Hence, one of the distinguishing features of ecosystem-based management is its emphasis on protecting the productive potential of the system that produces resource flows, as opposed to protecting an individual species or stock as a resource. In many other respects, however, ecosystem-based management is not very different from what many fishery managers trained in marine ecology already do: paying attention to species interactions such as competition and predation, conservation of habitat, and protecting critical life history stages by closing nursery areas and spawning locations to fishing. Another point to note is that if the ecosystem is already degraded, sustainability no longer makes sense as a goal. There is no point in maintaining an ecosystem which is impaired. Instead, the goal should be "rebuilding" or restoring the ecosystem (Pitcher and Pauly 1998).

The ecosystem approach and principles such as the ones outlined above, highlight several additional points which are crucial for the task of the manager. These are related to ecosystem complexity, controllability, predictability, risk, and uncertainty.

2.2.1 Ecosystem complexity: Implications for control and prediction

In his book on fisheries management, Anthony Charles (2000) refers to the "illusion of certainty" and the "fallacy of controllability." There are good reasons to think that the world is *not* predictable and controllable. But our conventional philosophy of management follows a tradition of positivistic science which assumes that it is and which is based on equilibrium thinking since Newton. But as the Panel points out, our ability to actually predict ecosystem behaviour is limited, and models based on equilibrium thinking often do not work. This is not only because we lack data; it is also because ecosystems are intrinsically and fundamentally unpredictable (Holling *et al.* 1995). Chaos theory teaches that many phenomena do not follow a simple, cause-effect logic — they are nonlinear and hence unpredictable. Tiny changes in one variable can have unforeseeable consequences in the larger system ("the butterfly effect"). We can never possess more than an approximate knowledge, and our ability to predict the behaviour of a multi-equilibrium complex system, such as the ocean, is severely limited.

In these complex systems, aquatic ecosystems in our case, system processes in fact seem to be nonlinear, and tend to be characterized by discontinuities, thresholds, and sudden changes ("flips"). As the Panel observes, once thresholds are exceeded, and an ecosystem shifts to a new state, such changes can be irreversible. A case in point is the 1970s case of the Peruvian anchovy, which did not decline gradually but collapsed in a discontinuous manner. Once the anchovy population collapsed, it did not recover gradually but was replaced by other species which, in turn, prevented it from regaining dominance.

Such experiences have influenced the way in which we view ecosystems. Further, ecosystem changes are not limited to the effects of humans. Ecosystems change with time, as the Panel observes, and many of these changes are of natural origin. Time-series of data have informed marine scientists and fishery managers that there can be marked differences in year-class strength of fish, in addition to random fluctuations and what appear to be 10 to 30 year "regime shifts" (Steele 1998). Large, infrequent disturbances (LIDs), such as major hurricanes, are a feature of all ecosystems (Turner and Dale 1998). There is accumulating evidence that many aquatic ecosystems have multiple equilibria. Some of these become apparent due to human-induced changes such as overfishing and nutrient-loading, and some due to natural processes. Thus, it has become very difficult to talk about "the balance of nature" or "the equilibrium of the ecosystem." Often, there are many. Hence, many ecosystem scholars no longer talk about the notions of stability (in the sense of a system having one equilibrium point) or resilience (to mean the ability of a system to bounce back to that equilibrium point — because there is none).

Instead, some scholars are defining the "resilience" of an ecosystem as its ability to buffer or absorb perturbations. For some, resilience defined in this manner is a key system characteristic. It refers to the capacity of the system for adaptive change. Hence, managing for resilience means accepting and working with natural variability and conserving the adaptive capacity of the system (Holling *et al.* 1995). This approach is fundamentally different from command-and-control style of management, which typically tries to reduce environmental variability (or ignores it, as in conventional MSY calculations) in an effort to produce a steady and predictable stream of economic benefits (Holling and Meffe 1996).

Thus, the emerging view of ecosystems emphasizes unpredictability (as opposed to predictability), multiple equilibria (as opposed to single equilibrium), resilience (as opposed to stability), threshold effects (as opposed to smooth changes), non-linear (as opposed to linear) processes, and the multiple scales in which these processes occur. These changes indicate a view of ecosystems that is much more complex than the view on which our conventional management approaches are based. Thus, the shift in the ecosystem paradigm has major implications for fishery management approaches. For example, once we recognize the limits of predictability of future yields of a given stock, then we also recognize the limits of fishery management systems based on sustainable yields.

Although there is still some debate among ecologists, many now hold that "ecosystems are not only more complex than we think, they are more complex than we can think!" Populations and social systems can also be similarly complex. As Holling *et al.* (1995) put it, we need to learn to live with uncertainty. Once we reject the Age of Enlightenment idea of "controlling" nature, then we can then come to terms, as many generations of ancient fishing cultures have, that we can manage resources through a learning-by-doing approach (see the Adaptive Management section later in this chapter). This does not mean rejecting science, but recognizing the limits of knowability and appreciating the knowledge held by fishers themselves.

2.3 Uncertainty and Risk

One practical implication of the complex systems view for the resource manager is the question of how to deal with uncertainty. Increasingly recognized as an issue to address, uncertainty reflects the probability that a particular estimate, piece of advice, or management action may be incorrect. Risk is the potential cost, in terms of societal benefits, of adopting the estimate, advice, or management action should it turn out to be incorrect. Precautionary measures are those used to reduce risk in the face of uncertainty.

Fisheries scientists have tried to deal with uncertainty in several ways. One way has been to attempt to quantify the variability in factors that contribute to uncertainty. This approach leads to explicit incorporation of uncertainty and risk in the conventional management models (Smith 1993), and often requires even greater amounts of data, information, and technical expertise than are required for the basic models. Another way of dealing with uncertainty has been to assume that it cannot be quantified or avoided, and instead to adopt a precautionary approach with ample margins for error.

Recently, there has been a call to develop and use fishery management models that are less precise than the traditional ones but with a focus on more responsive control systems (Walters 1998; Charles 1998a). In Chapter 1, we referred to the "broad brush" approach, as opposed to the "narrow arrow." Both the precautionary approach and the responsive control system approach are much less of an exact science than the old one was thought to be. Therefore, they must be characterized by process, transparency, participation, agreement, documentation, feedback, accountability, evaluation, and responsiveness. They could be termed "new fisheries management." Such new fisheries management is not only for small stocks; the dramatic failures with large and highly valuable stocks suggest that there is a need to think in similar terms for them also.

Several sources of uncertainty may affect the fishery management process. According to one classification, these include (a) randomness (stochasticity or "noise" in the system), (b) structural uncertainties, which are fundamental uncertainties reflecting ignorance about the nature of the system and which include chaotic behaviour inherent to complex systems in which a small change in one variable can affect the outcome, and (c) state and parameter uncertainties due to imprecise parameter estimates and unknown states of nature (Charles 2000). In addition to these items, there is also uncertainty due to implementation error, which is probably the most important type of uncertainty in fisheries (see **Box 6.1**).

2.3.1 The precautionary approach

The precautionary approach to environment and natural resource management made its global debut at the UN Conference on Environment and Development (UNCED) in Brazil in 1992. Principle 15 of the Rio Declaration states that

> *In order to protect the environment, the* Precautionary *approach shall be widely applied by* States *according to their capabilities. Where there are threats of serious or irreversible damage, lack of full scientific certainty shall not be used as a reason for postponing cost-effective measures to prevent environmental degradation.*

The precautionary approach was adapted to fisheries by FAO and incorporated into the *Code of Conduct for Responsible Fisheries* (FAO 1995, Sections 6.5 and 7.5). In the *Technical Guidelines to the Code of Conduct*, the application of the precautionary approach to fisheries is developed in further detail (FAO 1996b). The precautionary approach is also prominent in the *Straddling Fish Stocks and Highly Migratory Fish Stocks Agreement* (United Nations 1995).

A key element of the precautionary approach for fisheries is that fishery management systems should err on the side of conservation, particularly when there is the chance of irreversible changes that may degrade the equity of future generations. Erring on the side of conservation means that there must be a shift in the burden of proof from the conserver having to prove that the proposed use of the resource will have long-term or irreversible effects, to the user having to prove that it will not (Charles 1998b).

The shift in burden of proof is a change that does have the potential for abuse. However, the Technical Guidelines emphasize that the precautionary approach should not be taken to imply that "no fishing can take place until all potential impacts have been assessed and found to be negligible." What it does require is that all fishing activities be subject to prior review and authorization; that there be a management plan for each fishery that clearly specifies management objectives; and that the means of monitoring, assessing and controlling fishing impacts be clearly stated (FAO 1996b).

In the face of uncertainty, which is always the case in fisheries, and with the need to be responsible by exercising precaution, fishery management systems must be able to cope with a great deal of subjectivity, at least until there is a good scientific basis for management. We have argued that for small fisheries, providing a scientific basis may not be feasible or affordable.Therefore, for these fisheries, management systems that can deal with subjectivity are a basic requirement. These management systems will have to be characterized by clearly specified processes, means of communication among stakeholders that are oriented to reaching agreement on appropriate measures, transparency in the processes and decisions, and clear documentation. These are system characteristics that will:

- Lead to decisions based on the best available information;
- Allow stakeholders to buy into the management system;
- Promote the capacity of the system to have a memory, to learn, and to build on experience.

2.4 Protected Areas

Of the various types of protected areas for fisheries, the marine protected area (MPA) or its freshwater equivalent, is the most common (Salm and Clark 2000). A marine protected area is a spatially defined area in which all populations are free of exploitation (NRC 1999). The primary purpose of MPAs is to protect target species from exploitation in order to allow their populations to recover. Perhaps more important, MPAs can protect entire ecosystems by conserving multiple species and critical habitats such as spawning areas and nursery beds. Stocks inside these areas can serve as a "bank account" or insurance against fluctuations in and the depletions of populations outside the protected area caused by mismanagement or natural variability. Thus, MPAs serve an important role in the conservation of marine biodiversity. But they are also potentially important for fisheries. A great deal of international interest has followed from the discovery, mainly in tropical areas, that MPAs can lead to striking increases in the number and size of fish in the protected populations.

A compilation by the US National Research Council (NRC 1999) showed 13 studies of MPAs with statistically significant positive effects. They included examples from North, South, and Central America, Africa, and Asia. Some had been in operation for as long as 20 years and others for as little as two. Most covered fish species and some covered invertebrates. However, only seven of these studies compared target species *before* and *after* reserve establishment. The others compared populations inside and outside reserves, which is not as strong a control. Nevertheless, overwhelming differences between some of these inside and outside comparisons make it clear that the results are not accidental.

Well planned studies of MPAs are still needed to understand how protected areas work. We do know that MPAs do not always result in higher populations of desired species or in higher biodiversity. Some of these results have been explained by species interactions within the MPA, or in terms of extensive and irreversible degradation of the area before the establishment of a reserve. Certainly, MPAs do help protect ecosystem structure and function in general. The size of the MPA relative to home range and habitat requirements of target species is important. As well, it is clear that MPAs are effective in protecting species which are sedentary or have a limited range, as in many reef fish. For species that have a large range or highly mobile life history stages (such as planktonic larvae), MPAs can serve to protect the spawning ground, spawning aggregations or the nursery area. In other cases, protecting vulnerable life history stages of the adults, such as spawning migrations, may prove effective.

In some circles, MPAs have come to be advocated as *the solution* for all fisheries and ecosystem management problems. But, in fact, there are controversial aspects of MPAs. In many cases, the establishment of a MPA seems to be a necessary but an insufficient condition to meet conservation objectives. As well, the enforcement of MPAs, and their local acceptability appear to be key. These and other issues are discussed in more detail in section 6.3.4.

2.5 Adaptive Management

All fishery management systems learn from their successes and failures. Adaptive management goes one step further and relies on systematic feedback learning. Adaptive management is a relatively new approach in resource management science, but its common-sense logic that emphasizes learning-by-doing and its elimination of the barrier between research and management, resembles resource management systems based on ancient wisdom. Both rely on feedback and learning, and on the progressive accumulation of knowledge, often over many generations in the case of traditional systems. Adaptive management has the advantage of systematic experimentation and the incorporation of scientific research into the overall management scheme. Active learning in adaptive management deliberately attempts to accelerate the learning process by "probing" the fishery system experimentally (Walters 1986). In some cases, management policies may be used as "experiments" from which one can learn. In other cases, it is not possible or feasible to experiment with the fishery. In such cases, simulation models have been used to carry out "experiments".

Resource management, as a branch of applied ecology, is a difficult field in which to carry out scientific research. The difficulty is easy enough to explain:

"experiments take longer, replication, control, and randomness are harder to achieve, and ecological systems have the nasty habit of changing over time" (Hilborn and Ludwig 1993). The problem is not the inherent complexity of the system under study. Single cells are very complex systems too, and yet research progress in molecular biology has been spectacular in providing applications based on predictive models. By comparison, predictive models in ecology are hard to come by, and this is certainly true for fisheries management. One explanation for the dearth of predictive models is the propensity of ecosystems to change over time in an unpredictable manner, as touched upon earlier in this chapter.

Dealing with the unpredictable interactions between people and ecosystems as they interact, adaptive management takes the view that resource management polices can be treated as "experiments" from which managers can learn (Holling 1978; Walters 1986). Organizations and institutions can "learn" as individuals do, and hence adaptive management is based on social and institutional learning. Adaptive management differs from the conventional practice of resource management by emphasizing the importance of feedbacks from the environment in shaping policy, followed by further systematic (i.e. non-random) experimentation to shape subsequent policy, and so on. The process is iterative and based on *feedback* learning. It is co-evolutionary in the sense that it involves two-way feedback between management policy and the state of the resource. Hence, adaptive management is an inductive approach, relying on comparative studies that combine ecological theories with observation and with active human interventions in nature and with an understanding of human response processes, all in the context of purposeful learning (Gunderson *et al.* 1995).

Individuals commonly learn from mistakes; this also seems to be a way in which resource management institutions learn. The important point is, effective learning occurs not only on the basis of management successes but also failures. The international experience with environmental management agencies shows that there often is institutional learning following a crisis (Gunderson *et al.* 1995). Adaptive management recognizes that resource crises and management mistakes can be useful because they create learning opportunities.

Learning from mistakes presupposes that what is learned can also be remembered. The mechanism for institutional learning, like any learning, is trial-and-error. There are several reasons why management institutions may be slow and sporadic learners. The essential steps in learning from experience include documenting decisions, evaluating results, and responding to evaluation (Hilborn 1992). But even if these are done, management agencies have few mechanisms of institutional memory to retain the lessons learned. Publications, data records, and computer databases, even where they exist, are often not adequate to serve the institutional memory. The richest form of memory in an agency is stored in the brain of fisheries managers and field personnel. A smart manager learns to tap this knowledge, along with the knowledge of fishers.

Of particular importance is the memory of environmental fluctuations. We have already discussed the significance of decadal-scale regime shifts in marine ecosystems

(Steele 1998), and large, infrequent disturbances or LIDs (Turner and Dale 1998). What mechanism do we have to respond to such perturbations? Some government agencies keep records and maintain disaster-response plans. But the fisherfolk themselves may have a role to play as well. There is evidence that some traditional societies, such as those in the Western Pacific, maintain institutional memory of such LIDs as major tropical hurricanes, along with recipes of responses. Perhaps a combination of agency institutional memory and elder fishers' knowledge can help provide adaptive responses for ecosystem perturbations.

2.6 Management in Information Deficient Situations

In this book we speak in terms of fishery information rather than data. Although this may seem to be a semantic difference, we do so to emphasize the point that much valuable information input to management of small-scale fisheries may be, or may appear to be, qualitative or anecdotal. Thus it may not fit the notion of data in conventional fisheries science. Schemes can be devised to quantify this type of information, but this requires special attention. We look at this in greater detail in Chapters 4 and 5.

Over the past century, much progress has been made in the scientific study of fisheries, marine ecology, and oceanography. Yet despite the accumulation of a great deal of scientific data, there is insufficient information to manage fish stocks, especially those of multispecies fisheries in tropical seas. We have long been taught to believe that fisheries management requires extensive research, sophisticated models, large amounts of data, and highly trained experts. We now know that these ingredients do not always work, and are coming to realize that simpler approaches can be more practicable and cost-efficient, as we must "reinvent fisheries management" (Pitcher *et al.* 1998). In the case of small-scale fisheries, the costs of information and expertise are prohibitive. It is evident that we must reorient our thinking toward fisheries management systems that can work with much lower inputs of data and information, systems that can be developed and implemented by generalists with a broad working knowledge of the fisheries.

The case for management of small-scale fisheries in information-limited situations has emerged at various times in the past in several forms. Indeed it has been incorporated into the *Law of the Sea* and other international agreements, which state that management should be based on the best available scientific information, and should not be delayed due to inadequate information. These are now principles of the *Code of Conduct for Responsible Fisheries*. Recognition of the problems due to lack of information and the need for solutions that include short cut methods, commonsense approaches, stakeholder information, and consensus have been proposed by Caddy and Bazigos (1985); Mahon (1990, 1997); Johannes (1998a); Caddy (1999) and McConney (1998); mahon (1990, 1997); Johannes (1998a) provides several examples in which the use of local knowledge and commonsense led to improved management systems. He takes care to point out that such "dataless management" does not mean management without information, and emphasizes two sources of information that should be more widely and systematically used. The first of these is the traditional knowledge of the fishers. Even where there is no traditional knowledge, the stakeholders inevitably have information and ideas about the measures that could improve the fishery. The second source of information, according to Johannes, is the use of studies on similar fisheries in other locations. Given the various uncertainties, it may be more practical than is usually thought to use information from other fisheries, with a suitable safety margin. Improvement in availability and user friendliness of information systems can facilitate the process. However, it is up to the manager to build into the planning process a systematic search for relevant information from other fisheries that are useable in terms of resource types, technology, and human organization.

There is a need to access and use all possible sources of information, including information from comparable regions. One of the challenges faced by managers of small-scale fisheries is to look beyond the scientific paradigm and learn how to access information that is readily available and relatively inexpensive to collect. In this book, we deal with several kinds of such information. Chapter 4 explains how judicious observations of the fishery and the resource environment can yield "rapid appraisal" information. Information obtained by tapping the observations and understanding of the fishers themselves is analogous to rapid appraisal, and is usually referred to as traditional ecological knowledge, indigenous knowledge, or local knowledge.

2.6.1 Traditional ecological knowledge

Traditional ecological knowledge may be defined as "a cumulative body of knowledge, practice and belief, evolving by adaptive processes and handed down through generations by cultural transmission, about the relationship of living beings (including humans) with one another and with their environment" (Berkes 1999, p. 8). Traditional ecological knowledge is both cumulative and dynamic, building on experience and adapting to changes. It is an attribute of societies with historical continuity in resource use in a particular environment. Practical knowledge that does not have such historical and multigenerational character, can simply be called, local knowledge. This is recent knowledge, as in the non-traditional knowledge of some Caribbean region peoples (e.g. Gomes *et al.* 1998). Another term used in the literature, indigenous knowledge (IK), is more broadly defined as the local knowledge held by indigenous peoples or local knowledge unique to a given culture or society.

Local knowledge held by fishers may be about ecology, climate, and weather, technology, business, illegal activities, international trade and so on. Traditional knowledge, which can be seen as a special case of local knowledge, may pertain to biology and ecology, and it can also be about institutional organization and indigenous management systems. Much of the traditional knowledge literature is based on indigenous peoples. However, many non-indigenous groups, such as inshore cod fishers of Newfoundland (Neis 1992; Neis *et al.* 1996), no doubt also hold traditional ecological knowledge.

The current interest in the use of traditional ecological knowledge goes back to the early 1980s, following the documentation of the amazingly detailed knowledge held by the fishers of Palau, Micronesia, in the Pacific (Johannes 1981). Is traditional knowledge relevant to modern management? A number of suggestions have been made to include traditional knowledge in contemporary management systems (Dyer and McGoodwin 1994). Over the years, increasingly important roles have been proposed for traditional knowledge. For example, there is an emerging consensus in Oceania that, given the scarcity of scientific knowledge, alternative coastal fishery management models may be developed in which local knowledge may substitute for scientific data (Hunt 1997; Johannes 1998a). Such models have yet to be tested, but the question is timely: How can resource management be improved by supplementing scientific data with local and traditional knowledge? How can information from resource users themselves broaden the base of knowledge necessary for sustainable resource use?

Figure 2.2 illustrates the spectrum of local and traditional knowledge, and the spectrum of scientific and technical knowledge interacting with and enriching one another. In some cases, these two knowledge systems may be distinct. The traditional knowledge system may be based on a world view of human-nature relations that is very different from the Western one. Such is the case, for example, with the Cree Indians, an indigenous people of Canada (Berkes 1999). In those situations, it is extremely difficult, and perhaps inadvisable, to combine or synthesize the two kinds of knowledge, even though they still can inform one another, as shown in **Figure 2.2**. In other cases, for example in the Caribbean where distinct traditional systems do not exist, it is easier to meld science and local knowledge. Models that combine practical information and scientific knowledge are beginning to appear, as discussed further in section 6.2.6.

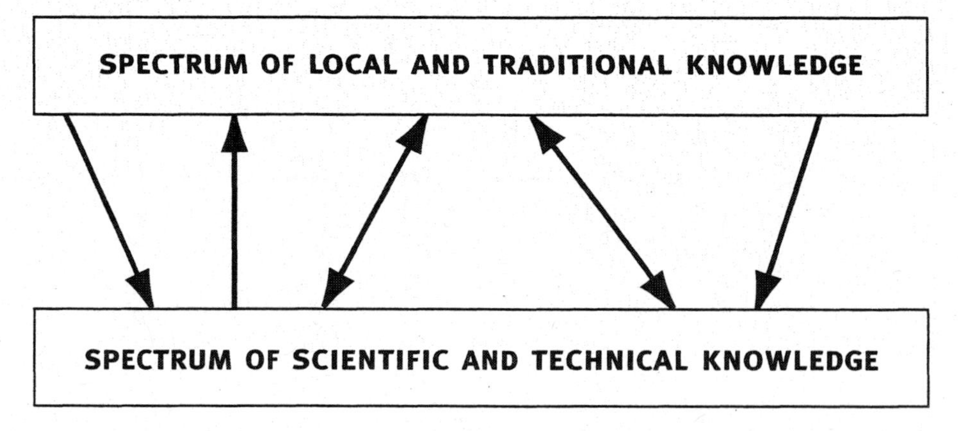

Figure 2.2 Exchange of information between knowledge systems.

The framework for a new fisheries management will have to be such that it can accommodate traditional knowledge, qualitative indicators, and proximate variables as means of evaluating the status of a fishery and determining future directions. The use of such information in the absence of other or better data is precautionary. In the absence of hard scientific evidence, when all stakeholders know and agree that a fishery is in an undesirable state, there should be no need to invest in research to provide the evidence before taking steps to move the fishery in the direction of improvement within a planning process (Chapter 3).

The ability to improve a fishery will be considerably strengthened when the stakeholders can agree on some measures to effect change. The key element at this stage in the process is agreement or consensus (Caddy and Mahon 1995). Thus, achieving consensus will be a large component of participatory management that is based on qualitative or traditional knowledge. It may be acceptable, even desirable, to approach management through simple rational schemes that can be understood by all of the participants. We will return to this management topic in Chapter 6, after taking a closer look at information in chapters 4 and 5.

2.7 Governance Regimes

Fisheries governance used to be simple. It has become complicated over the years, for a number of reasons. We have already discussed the changing scientific views of ecosystems, including the issue of management with incomplete information and the necessity of using fishers' knowledge. But there are other considerations as well. According to Chakalall *et al.* (1998), there are two main groups of issues in fisheries governance. The first is the need to strengthen national and regional fisheries organizations, to improve the management of shared stocks and to participate in international management initiatives. The second is the need for fisheries administrators to develop partnerships with non-governmental organizations (NGOs) in general, and fisherfolk organizations in particular.

The *United Nations Convention on the Law of the Sea of 1982* (UNCLOS) gives coastal states the authority to manage fisheries within their jurisdiction. As a result, most coastal states have had to revise their fisheries legislation. State jurisdictions have, been extended seaward, and, in many cases, provisions have been made to enter into agreements with other countries in the region and with regional organizations for cooperation in fisheries management.

With the emergence of extended fisheries jurisdiction under UNCLOS, the coordination of fisheries policy among a number of governmental bodies concerned with foreign affairs, shipping, tourism, economic development, cooperatives, marketing, and so on, becomes more critical. The overlap in responsibility is usually greatest between fisheries and the departments responsible for environment, tourism, ports, health and agriculture. These overlaps have been recognized in coastal area management or coastal zone management, a rapidly growing field internationally. Ultimately, fisheries must be managed within the broader context of coastal areas and the many factors that affect aquatic environments.

In many of the smaller developing nations, and especially in island states, many stocks are shared. But the issue is not confined to such countries. Even the larger developed nations, such as Canada, have found that large and economically important stocks are at the mercy of international fleets. Regional and international arrangements for the management of shared, migratory, and straddling stocks vary according to the distribution of the resource. Where only a few countries are involved, bilateral or multilateral arrangements are often used. For example, Trinidad and Tobago, Grenada, Barbados, St. Lucia, Dominica, and Martinique can cooperatively manage flyingfish on the basis of a multilateral agreement (Chakalall *et al.* 1998). For stocks that range more widely — for example, through the Caribbean region and into the High Seas — multilateral agreements will be needed with many more nations.

Turning to the second group of issues, fishery administrators need to develop partnerships with a range of organizations to improve management. These include a host of non-governmental organizations with an interest in fisheries or those groups whose activities impinge on fisheries. Various NGOs may have interests that coincide or conflict with those of fishers. Many environmental organizations work for the protection of the health of coastal areas and help conserve aquatic resources. For example, the Community Environmental Resource Centre (CERC) of Jamaica was involved in a grassroots environmental sensitization strategy by showing the relationship between improper sewage disposal and fouling of oyster beds and recreational beaches in the Kingston harbour (Chakalall *et al.* 1998).

However, NGOs that represent competing commercial uses of coastal areas may have interests that conflict with those of fishers (Brown and Pomeroy 1999). Examples include water sports and tourism associations — and these tend to have more political power than fisher groups. By lobbying for restrictions to commercial or artisanal fish harvests, tourism associations and diving groups may have a major impact on fishing communities and on resource management in general (Renard 1991). Can commercial fishing, recreational fishing, diving, tourist beaches, and boat harbours co-exist in the coastal zone? The integration of fisheries into coastal zone management is set out explicitly in the Code of *Conduct for Responsible Fisheries* (FAO 1995).

One group of organizations are particularly important for potential partnerships with government agencies: fisherfolk groups. Our discussion turns to a consideration of these organizations in the context of stakeholder groups and public participation in general.

2.8 Stakeholder Participation

An equation used by the Canadian International Development Agency (CIDA), succinctly summarizes the key elements of successful participation:

> ***The degree of successful participation = will + skill + organisation***

To achieve the ***will*** for participation, both government agencies and stakeholder groups may need to shift their perceptions about the role of participation in achieving results in fishery management. Similarly, capacity building is inevitably needed for both parties to build the ***skills*** to take part in the process constructively, and to develop the ***organizational*** platform from which to take part in the process. A recurrent theme of this book is the importance of the participation and empowerment of the stakeholders to the fullest extent feasible. This is not a philosophical position, although participation can be defended ethically on grounds of democratic principles — people whose livelihoods are potentially affected by a decision should have a say in how that decision is made. Rather, the point is that participation is also important for the effectiveness of resource management.

Citizen action, empowerment, stakeholder participation, civil society involvement in state management: whatever the name, this emerging global trend is having an impact on all aspects of public management and development at local, national, and international levels (Burbidge 1997). This trend manifests itself in many ways, for example, in the implementation of programs aimed at making participation a fact of everyday life[1]; in the reorganization of national and international institutions to accommodate stakeholder participation; and in a rapidly growing technical literature on the subject.

In small-scale fisheries, the scope for civil society participation is great, perhaps more so than in many other sectors. This is so partly because small-scale fishers have been marginalized in the conventional top-down decision-making processes, and there is now a trend toward greater community orientation (Christie and White 1997). But this is also due to the very nature of small-scale fisheries, which are virtually unmanageable without the input and cooperation of stakeholders. The use of imperfect information for management necessitates a close cooperation and risk-sharing between the management agency and the fisherfolk. Such a process requires collaboration, transparency, and accountability, so that a learning environment can be created and management can build on experience. Transparency means openness, and full and free availability of information, decisions, and plans. Accountability means the people who make the decisions should be available to answer to the people who are affected by the decision.

---

[1] For example the OAS ISP, Organisation of American States, Inter-American Strategy for Public Participation in Environment and Sustainable Development Decision-Making in the Americas (http://www.ispnet.org).

The first step in participation is the identification of stakeholders or actors. Who are the major players in the fishery? Who represents them? Are all actors represented? Is there legitimacy and representativeness — do the representatives speak for the members? Sorting out the actors and representation is referred to as stakeholder analysis (details in Chapter 3). This is an essential step in the management of a fishery because a manager needs to know the fishery, but also because the manager needs to have in hand the full line-up of stakeholders for the purposes of consultation, cooperation, consensus-building, and conflict resolution. For example, consensus decision-making on a local overfishing problem will bring a number of actors around the table. A case of conflict between two gear groups may require negotiation between them, with or without mediation by the government agency or an independent third party. Such conflict resolution mechanisms, the flip side of consensus-building, require the presence of a functioning and representative organization or stakeholder body.

2.9 Comanagement and Empowerment

Participation by **resource users** and stakeholders in fisheries management can take many forms. It may range from consultation by government with these groups, to their having full responsibility for a fishery or management area. There are many levels between these two extremes of participation, such as the formation of fisheries advisory bodies with representation from various sub-sectors, or cooperation in planning and enforcement at the community level. These bodies may simply be referred to as advisory bodies, or they may be called multi-stakeholder bodies, round tables or comanagement bodies.

Comanagement is the sharing of power and responsibility between the state and resource user groups in the management of natural resources (Pinkerton 1989a). Although there is no total agreement in the literature on the classes of participation and the kinds of comanagement, there is a rich literature on the international experience with comanagement. A number of well-documented cases exist of giving fishers more say in resolving resource conflicts and managing local fisheries (Jentoft and McCay 1995). Chapter 8 distinguishes community-*centered* comanagement from other types.

Pinkerton (1989b) and Jentoft (1989) have listed the potential benefits of co-management to include community-based development; conflict management; and the decentralization of resource management. Resource users benefit from participating in management decisions that affect their livelihoods; government benefits from reduced challenge to its authority. Pinkerton further identifies seven resource management functions that co-management may enhance:

(1) Gathering of data;
(2) Making logistical decisions, such as who can harvest and when;
(3) Allocation decisions;
(4) Protection of the resource from environmental damage;

(5) Enforcement of regulations;
(6) Enhancement of long-term planning;
(7) More inclusive decision-making.

Not all types of comanagement arrangements fulfil all these functions. Throughout the book, we assume that it is desirable to aim for the strongest level of comanagement that is feasible. But, in fact, the spectrum of comanagement is wide. **Figure 2.3** shows the range of comanagement possibilities, from full community control (and minimal government involvement) at one extreme, to full government management at the other. There is a strong argument in the comanagement literature that the term comanagement should be reserved to situations in which there is a sharing of power and responsibility between the users of a resource and the government manager. Merely informing or consulting does not, constitute comanagement. Within the context of managing the commons (described in Chapter 7), these ideas are developed more fully in Chapter 8, which presents the preconditions for successful implementation of the various levels of comanagement. Regardless of the level of participation selected for a particular fishery, there are essential roles for both resource users and stakeholders, and government fisheries departments.

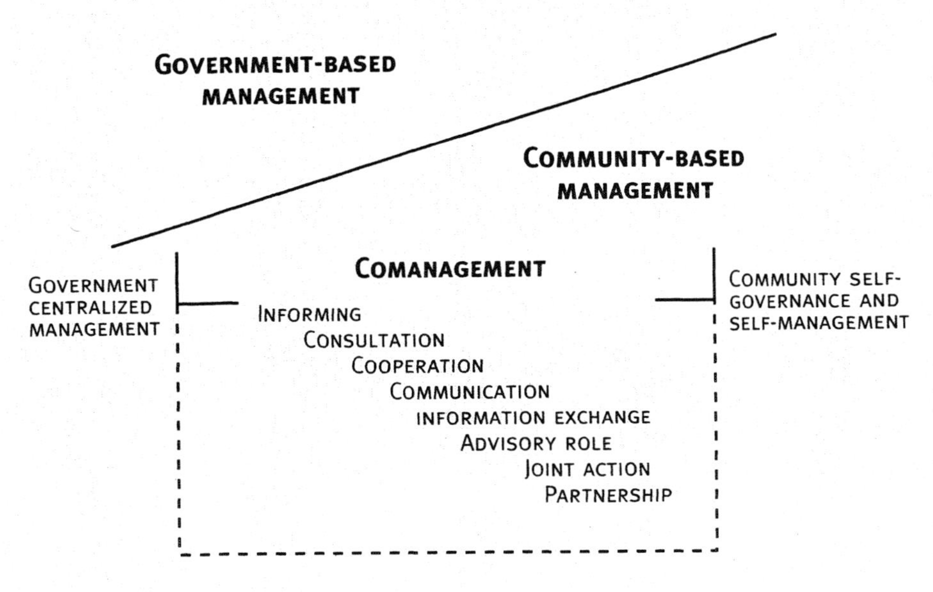

Figure 2.3 A hierarchy of comanagement arrangements (after Berkes 1994a).

For all parties to play these essential roles, there has to be empowerment. A group of fishers who have never had relations with government agencies, except as suspects of rule-breaking, will require some years of capacity building before they can settle into the new role as co-managers of a resource whose opinions are heard and respected. A community of fishers who do not even have the most rudimentary organization in terms of an association or a cooperative, and who have no civil society traditions such as neighbourhood associations, self-help groups, credit unions, or football clubs, will also require capacity building and empowerment.

The basic idea of empowerment is that a group of people have the power and the responsibility to get something done themselves. If a person or a group never has the power and the responsibility to make a decision or to share a decision, there is no empowerment. Empowerment proceeds in stages, and it is part of the skill and wisdom of a resource manager to nurture the organization of fishers, help the fishers develop confidence in themselves and trust in government, and, gradually, their capacity to solve problems. Such capacity building usually takes years and often involves non-governmental organizations working in close cooperation with both the managers and the users, as discussed further in Chapter 7.

2.10 Conclusion

This chapter introduced some of the key elements for a new kind of management approach for small-scale fisheries. We started with the changing concepts of ecosystem science, which emphasize unpredictability, multiple equilibria, resilience, threshold effects, non-linear processes, and the multiple scales in which these processes occur. These changes indicate a view of ecosystems that is much more complex than the view on which our current management approaches are based. This shift in the ecosystem view has major implications for the old fishery management paradigm, and provides the conceptual basis for a more holistic approach. Instead of assuming that we can control nature, we can come to terms with the limits of science, learn to use a diverse and creative set of information sources, start to use a learning-by-doing approach, and appreciate the knowledge held by fishers and other stakeholders.

Creative new management systems for combining traditional ecological knowledge from fishers and information from marine protected areas have been proposed—for example, for parts of Oceania — and are likely to be applicable to other areas as well. Management policies can be used as "experiments" from which managers can learn. These policy experiments potentially include the restoration of reef and lagoon systems in parts of Oceania. Adaptive management has two additional practical lessons for fishery managers: resource crises and management mistakes can be useful because they create learning experiences. Maintaining an institutional memory of these learning experiences is important.

Such considerations emphasize the importance of stakeholder participation, an emerging global trend. Given that fishery information will always be incomplete, cooperation and risk-sharing between managers and fisherfolk are essential. Collaboration, in turn, has to be based on principles of transparency and accountability. In many parts of the world, there has been an explosion of experiments in comanagement — the sharing of power and responsibility between the manager and the resource user (e.g. Brown and Pomeroy1999). Comanagement can provide for the empowerment of previously marginalized groups of fisherfolk. However, successful comanagement often requires years of capacity building before the parties can carry out their new roles in sharing management responsibilities.

The variety of concepts that pertain to fisheries is such that there are no straightforward formulas for management success. Instead, the management process requires that the value of the various approaches be evaluated case by case, and then integrated into a management strategy and management plan through consensus among stakeholders. As the conventional way of managing fisheries is gradually replaced by a new, more holistic and people-oriented way, there needs to be an emphasis on process. What makes the new management feasible is not a new formula that replaces the old, but putting into effect a new process of doing things. Chapter 3 considers what this process may look like.

Chapter 3
Fishery Management Planning and Objectives

3.1 Introduction

This chapter describes the use of planning and process in fisheries management. These are structured approaches for determining fishery objectives and selecting the best ways of achieving them. We set the stage for this by first outlining an approach to management of small-scale fisheries that takes into consideration the need to incorporate the wide variety of concepts and elements presented in the previous chapter. We emphasize process because much of the information is qualitative and there are not, as yet, and there will perhaps never be, any models that can combine this variety of information in a way that provides a recipe for management. Thus the manager's challenge is to balance, or even juggle, the various types of information and the interests of diverse stakeholders. We believe that achieving this will require a formal procedural framework as a point of reference for the stakeholders. Within such a framework, even in the absence of a comprehensive model, it will be possible to identify and implement measures to improve a fishery. In this process, the manager is the coordinator and facilitator.

The chapter considers fishery management planning, both the planning process and the plan itself. We distinguish between the planning process and the fishery management process, although the two are entwined (the book returns to the latter in Chapter 6). This chapter concludes with a look at the components of transparency and its importance in the management process.

One does not have to be a rocket scientist, stock assessment expert, and economist all rolled into one in order to go ahead and do something useful in the management of a fishery: listening to stakeholders, synthesizing their ideas, thinking about solutions, and coordinating go a long way.

3.2 An approach to management

This book focuses on the management of small-scale fisheries, the majority of which exploit small stocks. By small, we mean stocks with yields of less than 10 000 mt/year, although most stocks exploited by small-scale fisheries would probably produce yields of less than 1 000 mt/year (Mahon 1997). The amount of money that a country will be willing to spend on managing a fishery will likely be based mainly on its total value. Therefore, the value of a small-scale fishery rarely justifies the expenditure on data collection, analysis, and enforcement that large stocks do. The exceptions are where small-scale fisheries exploit large stocks, usually together with large-scale fisheries, or where the unit value of the resource is high enough that even though total yield is low, its monetary value attracts managers' attention. Lobsters, shrimps, and conch are examples.

Small stocks are as susceptible to overfishing as are large stocks, and are as much in need of management. For many developing countries, the majority of fishery yields come from a suite of small stocks. Failure to adequately manage this suite of resources can have a net or cumulative negative impact that is as high or higher than the collapse of a single large stock.

Because fisheries science has been developed mainly by scientists working on large stocks, which usually justify the cost of stock assessment, conventional fishery management is steeped in the need for data-intensive, biological assessment of the status of the resources. It has become almost doctrinal for managers to believe that little can be done until a stock has been assessed and management reference points chosen on the basis of that assessment. In that approach, which we describe as Stock Assessment Driven (SAD), management depends on monitoring the status through ongoing or periodic assessment. We suggest a different approach, one that is based on the view that even when a biological assessment is not affordable, there are usually viable alternatives. A great deal can be achieved with organization, planning and stakeholder participation: a Management Objective Driven (MOD) approach.

The most rational initial focus appears to be what the stakeholders want out of the fishery; that is, the management objectives. The focus can then shift to how to achieve the desired objective, and how to determine when the desired objective has been achieved; that is, how to measure successful management. This approach requires the knowledge and information to identify variables that relate to the objectives, followed by the setting of target points on those variables. Finally, we need control measures and systems that can be expected to bring about the desired changes.

This Management Objective Driven approach ensures that the management system focuses on the acquisition and analysis of data that relate to the objectives and control system. **Figure 3.1** contrasts the standard Stock Assessment Driven and Management Objective Driven and approachs. The SAD approach focuses on optimization or even maximization of the yield from the resource.

An advantage of the MOD approach is that it can be started with little or no quantitative information about the fishery. The process can and should be iterative. It can begin with broad objectives and simple short-term measures that will move the fishery in the direction of the objectives. It can incorporate obvious, common-sense improvements or controls. As information becomes available, the plan can be revisited and improved. This approach to management is consistent with the precautionary principle that is now embodied in most international agreements on fisheries and environmental conservation (FAO 1995, 1996b). It is also consistent with other elements of the international agreements that state that management should make the best possible use of the available information and should not be delayed while managers wait for better scientific information (United Nations 1992, 1995).

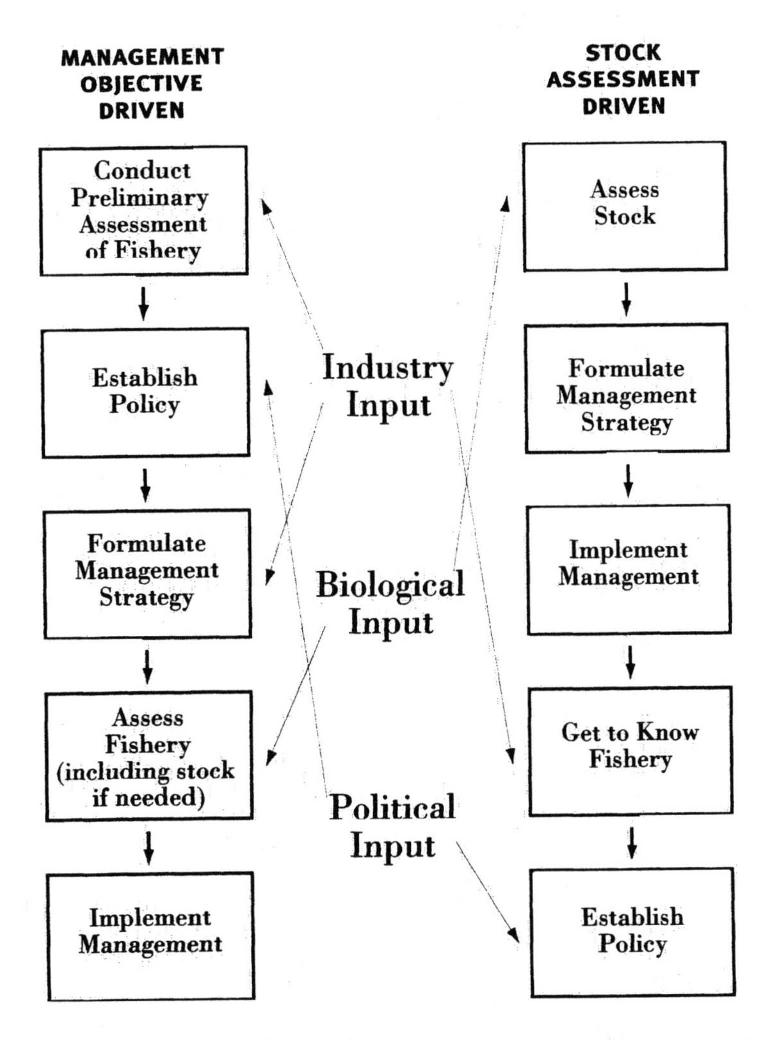

Figure 3 1 The action sequence that should take place when fishery management is management objective driven (MOD) and that tends to take place when it is stock assessment driven (SAD) Source: Mahon 1997

3 2 1 Investment in management

Above, we suggest that the level of investment in management is generally related to the worth of the resource, which may include non-fishery value such as biodiversity, culture, religion or ecosystem integrity While this is obvious to us, we have not found analyses that develop easily applicable formal or informal rules to guide the manager in determining an appropriate level of investment. Indeed few studies even quantify existing levels of investment (Arnason *et al.* 2000). For large-scale fisheries in developed countries, the costs of management may be substantial and variable, ranging from 3 percent of the value of the fishery in Iceland through 10 percent in Norway to 15 to 25 percent in Newfoundland, Canada (Arnason *et al.* 2000). This area needs further analysis by fisheries economists in order to give managers at least some rules of thumb about appropriate levels of investment.

It would be complex to quantify the costs of *not* managing in order to determine an appropriate level of expenditure on management, because leaving a fishery unmanaged can affect society and the economy in a variety of ways. The most obvious of these is the social and economic cost of the collapse of the fishery. This would include the cost of unemployment benefits for all those individuals who depend on the resource for a livelihood, including those who provide services to the fishery. In countries where such benefits are minimal, the incentive to invest in management may be low. This is even more the case where small-scale or artisanal fisheries are concerned, because the individuals affected are at the low end of the income scale and often live in remote areas. The fact that fishing may be part of a multi-occupational lifestyle may obscure the effects of overfishing, further undermining the incentive to invest in management. Indirect costs to be considered include health care to address malnourishment when protein supply is lacking and law enforcement when unemployed persons resort to crime.

Incentives for investment in management may be higher where the fishery has additional value to the country, such as providing foreign exchange, or where the fishery products are culturally important. In the latter case, the value placed on a fishery may be much higher than its measured economic value.

3.3 Management planning process

Although the fisheries literature frequently refers to the need for planning, it offers little guidance on how to set about planning the management of fisheries. For example, the FAO technical guidelines for responsible fisheries and their guidelines for data collection emphasize the importance of planning but do not go into detail on how to do it (FAO 1997, 1999). Hamlisch (1988) provides one useful attempt to do this, specifically for African countries. There is a considerable body of literature on economic and physical planning in general, but with a few exceptions from earlier times (for example, some FAO), the fisheries manager is left with the task of accessing and adapting the methods to his or her fishery's needs. The need to incorporate planning into training programs for fisheries managers is, however, gaining recognition (Msiska and Hersoug 1997).

Despite the fact that planning is a discipline, one can accomplish a great deal without formal training by taking a structured, common-sense approach to preparing a fishery management plan (FMP) based on stakeholders' knowledge and their understanding of the problems and potential of the fisheries. If the services of a trained planner are accessible, all the better. If not, one can still prepare an effective plan, since it is probably easier for a fisheries manager to gain a working knowledge of planning than for a planner not familiar with fisheries to immediately engage in fisheries planning.

At the national level, the FMP will be part of a sectoral plan for fisheries or for fisheries and agriculture. This sectoral plan will in turn be developed in the context of a macro-level economic plan. Aspects of the fisheries plan that pertain to land use and land-based physical infrastructure will also be related to the national physical development or environmental management plans. For planning to be effective, the linkages among the various types of plans that are used in each country must be considered to avoid conflicts and achieve positive, reinforcing interactions.

Most countries are home to several fisheries, all of which require some degree of attention from managers. Therefore, the fisheries management plan must address them all and consider the linkages among them, which include the effects that management of one fishery may have on another, and their common requirements for services or other inputs.

Ideally, the fisheries management plan (FMP) will:

- Represent the consensus of all stakeholders on how the fishery will be managed;
- Promote transparency by providing a clear statement, in terms that each stakeholder can understand, of what is expected to take place for each fishery, and, overall, within the time frame of the plan;
- Provide continuity when staff change in participating organizations;
- Provide a means of communicating the intentions and needs of fisheries to new stakeholders; for example, donors.

To achieve these aims, the plan must be documented in a form the stakeholders can understand, and they must have easy access to it.

3.3.1 The Barbados example

Figure 3.2 shows an example of a basic fisheries planning process, one that was used to develop an FMP for Barbados (McConney and Mahon 1998). This plan was mentioned in this book's introduction. The *Fisheries Act* of Barbados requires that the Chief Fisheries Officer develop and keep under review schemes for the management and development of fisheries, but it does not provide much detail on what should be in the plan or how it should be prepared. The initial formulation of the FMP for Barbados was based on a preliminary assessment of the fishery sector and the individual fisheries. In this case, it was prepared by the government Fisheries Division. The first stakeholder input took place through the Fisheries Advisory Committee (FAC), which the *Fisheries Act* also requires. At the next stage, all stakeholders were invited to review the draft FMP, and public hearings were held to facilitate that process. The plan was then approved at the Ministerial level and implementation began. The FMP will be reviewed and revised in the FAC every three years, following which the public will be provided the opportunity to comment on proposed changes.

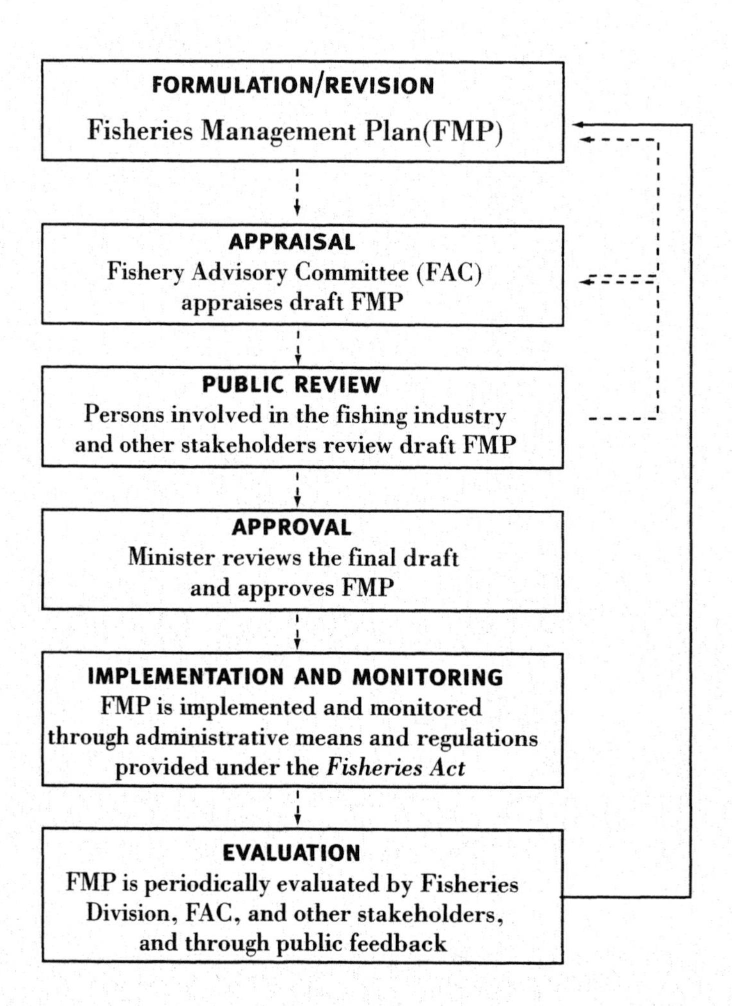

Figure 3.2 The process used to develop and review the Barbados FMP.
Source: Adapted from McConney and Mahon (1999)

The process shown in **Figure 3.2** provides for a moderate degree of stakeholder input. Representatives from the major stakeholder groups, the FAC members, contribute detailed input at an early stage. The wider public provides a consultative review rather than taking part in the formulation of the plan.

3.3.2 The US example

The USA provides an example of a country in which fishery management planning is very formal with considerable emphasis on the process and elaborate supporting institutional structure. *The Magnuson-Stevens Fishery Conservation and Management Act*, passed in 1976 and amended in 1996, specifies the process and content in great detail. This Act created eight regional fishery councils to manage the living marine resources of the waters of the USA and dependent territories. The councils' membership includes commercial and recreational fishers, marine scientists, and state and federal fisheries managers, who collaborate to prepare Fishery Management Plans (FMPs) for all exploited resources. The FMPs are prepared through a planning process that includes the public comments provided by fishers and other persons concerned with the management of these resources. There are six phases in the process:

- Phase I – Planning — in which the need for management is identified, a steering committee appointed, advisory panels formed, a plan development team appointed, and a work plan established;
- Phase II – Information gathering — in which all the relevant information on the fishery is compiled and the plan is drafted and reviewed internally by the Council committees and panels;
- Phase III – Review — in which the plan is subjected to a thorough technical review by the National Marine Fisheries Service, other relevant government agencies and public hearings;
- Phase IV – Formal secretarial review and implementation — in which the plan is officially reviewed and approved, and implementation begins;
- Phase V – Monitoring — in which the implementation of the plan is monitored;
- Phase VI – Revisions and amendments — in which any changes or new information are incorporated, as well as any emergency action that might be needed.

The contents of the plan are also specified in detail. Each plan must be complete and be supported by all the prescribed documents before it can be reviewed and implemented. The information that is required may call for extensive research and analysis. However, the process can proceed with the best available information. More information on the laws, the fishery management councils and the fisheries management process employed in the USA is available at http://www.nmfs.gov.

3.3.3 On planning for small fisheries

Clearly, the planning process that is prescribed by the law in the USA requires data, information, and expertise that would not usually be available, and perhaps not be appropriate, for many small stocks or small-scale fisheries. We outline their process here to illustrate the emphasis on process — the Councils were established to ensure that the process was followed. The process serves to ensure that all available information is considered, that the main issues are given the necessary attention, and that the stakeholders have had the opportunity to contribute to the process and to understand the implications of what is being proposed. We believe that even when there is limited information and expertise, the establishment of and adherence to a clearly stated process that involves stakeholders and uses the best available information in a reasonable way will significantly improve fisheries management for small-scale fisheries.

Returning to **Figure 3.2**, the first step is the formulation of a draft FMP. This is based on a preliminary assessment of the fishery and should assemble, in one document, all the available information on the fishery. The types of information that can be included in a preliminary fishery assessment and their uses are discussed below. Much of this information will be available as common knowledge among members of the fisheries department and industry stakeholders. It is important to retrieve and document this information so that it becomes available to all stakeholders and can either be a part of the consensus or can be identified as a subject for further clarification and discussion.

At this stage, depending on resources available, it may be desirable to carry out studies. However, we emphasize that the first draft should be based on readily available information, or information that can be acquired in a reasonable time frame; say, three months. The preliminary fishery assessment should identify the need for further information but should not wait on it.

Discussion about management objectives and strategy or approach, the core of the draft FMP, should be based on the preliminary assessment. In the unlikely event that there are no reasonable ideas about what needs to be done to improve the fishery while making it sustainable, it will be necessary to explore precautionary measures.

At this stage, the plan should be holistic. That is, it should be seen to include development (improvements in the fishery for the benefit of the industry stakeholders and consumers) as well as measures needed to conserve the resource. If industry stakeholders are to take part in the process of developing and implementing a plan, they will be more motivated if the plan includes improvements as well as control measures.

3.3.4 A PROJECT APPROACH TO PLANNING

One approach to planning is to view management as a project. Indeed, in many developing countries, progress in fisheries management takes place through a series of projects. Often, these projects take place with little or no wider context. The treatment of fisheries management as a project provides the opportunity to develop a suite of linked projects (a program) to address the fishery management needs of the country. The elements of the overall program can be managed as projects with discrete beginnings and endings. This approach also facilitates the acquisition of funding from donors and the national budget. Chapter 5 provides detail on the project cycle and process as they apply to fisheries management and improvement projects.

The value of approaching fisheries management planning as a project is the requirement that goal, purpose, objectives, activities, outputs, and means of evaluation be clearly stated. It also puts the exercise in a specific time frame. Logical Framework Analysis (LFA) is a popular approach to developing, communicating, and managing projects. Many donors require that projects submitted for funding be prepared in this format.

A full description of LFA would require more detail than can be provided here. Guides to the use of LFA in project planning have been developed by several agencies (Commission of the European Communities 1993; USAID 1994; IADB 1997). Here we provide a brief outline of the structured approach that LFA uses.

LFA methodology was developed in late 1979 and the early 1980s as a tool for the conceptualization, design, and execution of development projects. It consists of a series of processes:

- Stakeholder analysis
- Problem analysis
- Objective analysis
- Analysis of alternatives
- The logical framework matrix
- Execution plan
- Monitoring and evaluation plan
- Project reports.

Most of these steps are self explanatory, and are further explained by IADB (1997). The logical framework matrix is peculiar to LFA and requires further explanation. Its purpose is to summarize the project clearly and succinctly in a standard format. The rows of the matrix are referred to as the vertical logic. From the top down are: the goal that the project serves, the specific purpose of the project, the outputs that will be generated in order to achieve the purpose, and the activities that will be carried out to generate the outputs. The columns of the matrix, the horizontal logic, are: the objectives, the indicators that the objectives have been achieved, the means of verifying the indicators, and the assumptions upon which the achievement of the objectives is based. The matrix also summarizes the resources that would be needed to produce the outputs.

Despite the widespread use of LFA as a project planning and management tool, we were unable to find published examples of its use as a fisheries management planning tool. Therefore, we categorize it as a methodology that appears to be useful and should be explored. To further illustrate the potential of LFA, we have constructed one for the fisheries of Barbados (**Table 3.1**), focusing successively on the management subcomponent and, within that, the deep demersal fishery.

Despite the fact that most fisheries departments frequently deal with projects, staff are seldom trained in project development and management. A fisheries training course for South African Development Community Nations that included this material revealed that students could integrate course material on LFA with that on economics and management and planning but were less able to do so with course material on sociology, biology, and technology (Msiska and Hersoug 1997).

Table 3.1a Overall logical framework for fisheries management and development program in Barbados.

Goal of fisheries management and development: To ensure the optimum utilization of the fisheries resources of the waters of Barbados for the benefit of the people of Barbados		
Purpose of management: To address the specific and collective needs of the stakeholders in Barbados' fisheries		
Subprogram 1 Administration and Services	Subprogram 2 Fishery Industry Development	Subprogram 3 Fishery Management
Purpose: To provide the institutional and administrative basis for the development and management of the fishery resources of Barbados	**Purpose**: To develop and enhance the fishing industry of Barbados so that it can make optimal use of the fishery resources	**Purpose**: To conserve the fishery resources of Barbados so that they retain the capacity to provide optimum sustainable benefits for the people of Barbados
Outputs: • Well-trained and professional fisheries administrative staff • Administrative systems that meet the needs of fishing industry stakeholders in an efficient and cooperative manner • Capacity to develop and enhance the fishing industry • Capacity to develop and implement fishery and ecosystem management plans	**Outputs**: • Well-trained, professional fishing industry stakeholders • Well-designed and constructed fishing vessels appropriate to the needs of the fishery • Adequate shore-based infrastructure for servicing vessels, landing fish, and meeting the other needs of industry stakeholders • A reasonably priced, steady supply of good-quality fish to consumers • Active responsible fisherfolk organizations	**Outputs**: • Fishery-specific and ecosystem management plans • Supporting legislation • Implementation of plans • Fishery resources and ecosystems that are capable of long-term sustainable production of optimal yields

TABLE 3.1B LOGICAL FRAMEWORK FOR FISHERY MANAGEMENT SUBPROGRAM.

Goal of fishery management subprogram: To conserve the fishery resources of Barbados so that they retain the capacity to provide optimum sustainable benefits for the people of Barbados

Purpose of fishery management subprogram: To determine the optimal sustainable yield that can be produced by the resource and to devise and implement measures to ensure that those yields are obtained

FMP 1: SHALLOW-SHELF REEF FISHES	FMP 2: DEEP-SLOPE AND BANK REEF FISHES	FMP 3: COASTAL PELAGIC FISHES	FMP 4: LARGE PELAGIC FISHES	FMP 5: FLYINGFISH	FMP 6: SEA URCHINS	FMP 7: SEA TURTLES	FMP 8: LOBSTER
Purpose: Rebuild reef fish populations to levels capable of satisfying the requirements of both the commercial fishery and recreational or tourism non-harvest uses	**Purpose**: Ensure sustainable yield for local consumption, particularly in the off-season for pelagics, through a precautionary approach	**Purpose**: Optimize catches of the target species, particularly to meet the demand for input into other fisheries as bait, while minimizing bycatches of reef species	**Purpose**: Maximize catches within regional or international guidelines by ensuring fair and equitable distribution of these resources among the users	**Purpose**: Establish, in cooperation with other countries, a regime that facilitates long-term sustainability with an acceptably low risk of fishery disruption due to catch variability	**Purpose**: Rebuild populations and establish a comanagement arrangement with fishers	**Purpose**: Protect, conserve and rebuild sea turtle populations	**Purpose**: Ensure sustainable harvest of lobster for domestic and local tourism use to achieve the maximum long-term economic return from the resource
Outputs: • Reef fish populations recovered or recovering, with established protected areas for tourism industry to use	**Outputs**: • Steady supply of good-quality fish, particularly during the off-season for pelagics	**Outputs**: • Optimal catch levels identified, and measures in place to achieve these levels	**Outputs**: • For regional resources, agreement among fishing countries on management mechanism and equitable shares • For ocean-wide resources, participation in ICCAT to obtain good management and allocation of fair share	**Outputs**: • Agreement among fishing countries on management mechanism and equitable shares • Regional management plan developed and implemented	**Outputs**: • Comanagement mechanism established and operating	**Outputs**: • Internationally agreed measures to rebuild sea turtle populations in place	**Outputs**: • Steady supply of lobster to local consumers and tourism industry, based on agreed catch levels and measures to ensure these levels

Table 3.1c. Logical framework analysis — deep-slope and bank reef fishes.

	INTERVENTION LOGIC	OBJECTIVELY VERIFIABLE INDICATORS	SOURCES OF VERIFICATION	ASSUMPTIONS
OVERALL OBJECTIVES	Ensure sustainable yield for local consumption through a precautionary approach to further development	Comanagement mechanisms in place and resources being managed	Data on landings and reports from fishers	Stakeholders willing to cooperate
PROJECT PURPOSE	Steady supply of good-quality fish, particularly during the off-season for pelagics	Deep-slope and bank fish species available for purchase throughout the year	Fisheries Division record, interviews with retailers and restaurants	No major natural disturbances that disrupt production or negatively impact the resource
OUTPUTS	1. Fisherfolk participating in planning and management 2. Minimum mesh size for traps set and enforced 3. Protected areas established and enforced 4. Response of resource to precautionary measures monitored and evaluated	1. Mechanism for fisherfolk input established and operating 2. Agreed mesh size in regulations and few or no traps with smaller mesh in operation 3. Protected areas in regulations; little or no fishing taking place in them 4. Catches of resource stable or increasing	1. Interviews with fisherfolk organizations 2. Published regulations and reports from inspectors and fishers 3. Published regulations and reports from inspectors and fishers 4. Technical reports of monitoring and interviews with stakeholders	Fishers willing to participate Funds available for assessment

ACTIVITIES				PRECONDITIONS
	1. Locate fisherfolk involved in fishery and initiate dialogue 2. Determine appropriate precautionary mesh size 3. Seek agreement on mesh size 4. Draft and gazette mesh size regulations 5. Determine number and sizes of areas that would be appropriate for protected status 6. Seek agreement on areas 7. Draft and gazette regulations on areas 8. Establish monitoring program	**Human resources** List number of person years of individuals with various skills required to carry out this project component **Material resources** List all supplies, equipment, transportation, workshop, and other resources required to carry out this project component		

3.4 The management plan — what should it include?

A holistic approach to fishery management will lead the manager toward a fishery assessment rather than a stock assessment. When all the fisheries of a country are considered together, the fisheries assessment will become an exercise that is commonly referred to as a fishery sector review or assessment. The manager will usually perceive a sector review as a huge and daunting task that cannot be undertaken without external assistance with funding and expertise. The same general arguments for undertaking a preliminary fishery assessment apply equally to a sector review. It can be approached using available information and expertise, then reviewed and revised as information becomes available. Review and revision need not encompass the entire sector assessment: it can be broken down into its component parts, which can be tackled as the need arises and information becomes available.

The geographical scope of a sector review will depend on the size of the country and the linkages among fisheries. For small coastal and island states, the review will typically include the entire country. For larger countries, there may be little interaction or geographical overlap between freshwater and marine fisheries, or between fisheries on different coasts. The review can therefore be carried out in areas of the country within which there are interactions among the fisheries.

A fishery sector review, which typically covers the topics shown in **Table 3.2**, provides the information base for the fisheries development and management plan. The contents of a typical plan show the diversity of information that may be relevant to managing a fishery (**Table 3.3**).

At each step, there is the need for balance between the cost of information, the level of detail needed, and the value of the fishery. Much of the needed information is often already available, even in the fisheries department, but is not compiled and structured in a way that it can be used in a planning process.

The sector review and FMP can be as simple or as detailed as is warranted by the size and complexity of the fisheries and the amount of information available. In some cases, it may be possible to combine the sector review and plan. It may also be desirable to have two versions of the sector review and plan: a technical version that includes all the information in detail, and a public information version that includes the information in a form that the stakeholders can understood use. The Barbados Fisheries Management Plan, an example of the latter (**Table 3.4**), includes both sector review information and plans for the development and management of fisheries in a form suitable for dissemination to the public. (See Chapter 4 for more on fisheries information.)

TABLE 3.2 THE ELEMENTS OF A TYPICAL FISHERY SECTOR REVIEW.

REVIEW OF FISHING INDUSTRY OVERALL
Identification of individual fisheries that comprise the sector
Resource base for each fishery
Resource types
Habitats
Harvesting sector (including recreational)
Fleet
Fishers
Landing sites and infrastructure
Post-harvest sector
Processing
Retailing
Exports
Support services
Boat builders
Gear and equipment suppliers
REVIEW OF INSTITUTIONAL AND POLICY SUPPORT TO THE FISHING INDUSTRY
Legal and policy framework
Fisheries legislation and policy
Fisheries regulations
Related national legislation and policy, trade, environment, foreign
Linkages at the national level
International policy and agreements
Institutional fisheries management capacity
Fisheries department staffing and capacity
Fisherfolk organizations
FISHERY-SPECIFIC REVIEWS
Summary of biological knowledge
Local
Other relevant studies
Summary of management options
Local
Other relevant systems

Table 3.3 The contents of a basic fisheries management plan.

Fisheries management and development plan
Harvesting sector
Options for improvement of fleet
Options for shore-based facilities
Options for fishers, including organizations
Post-harvest sector
Options for processors
Options for retailers
Options for exporters
Data and information needs for planning
Preliminary valuation of sector
Institutional strengthening
Fisheries division
Fisherfolk organizations
Fishery-specific management plans

The plans produced by the US Fishery Management Councils are examples of plans that tend toward the fullest possible detail for all fisheries. Even so, these plans vary considerably in complexity and completeness, depending on the available information and the value of the fishery. The contents of a typical FMP produced by the US fisheries management planning process do not differ substantially from those of the simpler plan used in Barbados. What differs in most cases in the level of detail and analysis provided under each heading (**Table 3.5**). A notable recent addition to the US FMPs is the inclusion of an appraisal of essential fish habitat as required by the 1996 US *Sustainable Fisheries Act* (US SFA).

The SFA is also leading to the more explicit inclusion of ecosystem-based management in the plans as described in Chapter 2 (EPAP 1999). However, there is a proposal in the USA that a separate Fishery Ecosystem Plan (FEP) be prepared for each ecosystem. This approach will address some of the problems arising from the differences in spatial scale and complexity between individual fisheries and the fisheries ecosystem of which they are part (section 3.5).

Certain kinds of information are fundamental to successful management:

- Definition of the management unit biologically, socially, and spatially (section 3.5);
- Knowledge of the primary stakeholders: who, why and where;
- Understanding of the relationships between stakeholders and fishery (can be communicated using path diagrams for flows of products among stakeholders) (see Chapter 4);
- Review of related national policy so that fisheries can operate within a rational policy framework.

TABLE 3.4 THE CONTENTS OF THE FISHERIES MANAGEMENT PLAN FOR BARBADOS.

1. Guiding principles	5. Fishery-specific management plans
1.1 Mission	5.1 Shallow-shelf reef fishes
1.2 Goals of fisheries management and development	5.2 Deep-slope and bank reef fishes
1.3 Fisheries policy and plan	5.3 Coastal pelagics
1.4 Country profile	5.4 Large pelagics
2. Fishing industry profile	5.5 Flyingfish
2.1 Intersectoral linkages	5.6 Sea urchins
2.2 Overview of fisheries to be managed	5.7 Sea turtles
2.3 Fishing industry	5.8 Lobsters
3. Fisheries management	6. Fishery management options
3.1 Fisheries planning process	7. Glossary
3.2 Coastal zone management	**CONTENTS OF THE FISHERY-SPECIFIC MANAGEMENT PLANS.**
3.3 Fisheries-related legislation	Target species
3.4 Regional fishing	Bycatch
3.5 Organizational framework	Ecology
3.6 Fisheries research and statistics	Description of fishery
3.7 Fisheries monitoring, control, and surveillance	Management unit
3.8 Inspection, registration, and licensing systems	Resource status
4. Fisheries development	Catch and effort trends
4.1 Vision of harvest sector	Regulatory history
4.2 Vision of post-harvest sector	Management policies and objectives
4.3 Vision of state sector	Selected management approaches
	Development constraints
	Development opportunities

We cannot provide a comprehensive review of the types of information that could be relevant to the sector review and plan or of the methods that are available to acquire this information. We do want to communicate that it is up to the developer of the plan to include the information that is relevant to the objectives of the fishery in question. We also want to make the point that fisheries managers and researchers are increasingly attempting to define the goals and problems of fishery management in much broader terms than has previously been the case. In so doing, they are using a wider variety of information and are using simpler and more innovative methods to acquire this information.

TABLE 3.5 THE CONTENTS OF A FISHERY MANAGEMENT PLAN IN THE USA.

EXECUTIVE SUMMARY
DEFINITIONS
INTRODUCTION
Description of resource
The fishery management unit
Abundance and distribution
Reproduction and early development
Growth and maturation
Movement and migration
Description, distribution, and use of essential fish habitat
Food
Predation
DESCRIPTION OF FISHERY
History of exploitation
Processing and marketing
Current status of the fishery
Florida management program
Catch and capacity descriptors
PROBLEMS IN THE FISHERY
Overfishing
Management/enforcement
Database
Information/education
Threats to essential fish habitat
MANAGEMENT OBJECTIVES
Overfishing definition
Rebuilding program
MANAGEMENT MEASURES AND ALTERNATIVES
Proposed measures
Other measures considered and rejected
Procedure for adjusting management measures
Essential fish habitat conservation recommendations
RECOMMENDATIONS TO LOCAL GOVERNMENTS AND OTHER AGENCIES
RELATED MANAGEMENT JURISDICTIONS, LAWS, AND POLICIES
Federal
Local
ESSENTIAL FISH HABITAT RESEARCH NEEDS

3.5 THE FISHERY MANAGEMENT UNIT

A successful plan requires a clear statement of the entity that is to be managed — the management unit. Management focused on stock assessment takes the unit stock as the management unit. Much has been written about the definition and identification of stocks. The ideas pertaining to stocks are closely related to definitions of population, subpopulations and the extent of interbreeding among these units. Studies of genetic relatedness among these units are often brought into play, and there is frequent discussion about how much gene flow between populations there can be and the units still be treated as unit stocks. Concepts and research relating to the discreteness of stocks and populations are highly relevant to fishery management, but for small-scale fisheries, this type of information will frequently be unavailable, necessitating more practical definitions of the management unit.

Clearly, if management addresses only one part of a large resource that is being affected by heavy exploitation in other areas, its chances for success will be constrained by those outside forces. Consequently, there is a need to define management units within which there is the greatest chance for success. Here the precautionary principle has a role. In the absence of good information on the extent of a stock, it is precautionary to use the largest feasible management unit. In this context, management should not be confused with local efforts at fishery improvement in communities. The latter efforts can be successful for subunits of the management unit, provided they do not depend on a response from the entire resource.

Ideally, the fishery management unit will encompass the entire resource and all of the vessel and gear combinations that exploit that resource. In many cases, where the resources are shared, this will require international cooperation for managing them. The *Law of the Sea*, and in particular the elaboration for highly migratory stocks and straddling stocks, provides a wealth of information on approaches to the management of shared resources (United Nations 1995). In many cases, although the scientific research may not be available to provide unequivocal definitions of management units, stakeholder knowledge can be used to reach consensus on reasonable units for management. For example, at an expert consultation on shared resources, fishery managers from the eastern Caribbean were able to agree on which species could be managed at the national, subregional, and ocean-wide levels (Mahon 1987). For some resources, feasibility and practicality were key ingredients in defining the management units. These were reef-related, demersal species (for example, reef fishes and lobster) that are known to have planktonic early life history stages that can disperse from one country to another. Despite participants' recognition that dispersal could result in recruitment linkages among countries, they agreed that, for practical reasons, individual countries should proceed to manage these resources at the national level as if they were independent management units. In contrast, the participants agreed that for pelagic species known to move between countries as adults, management should encompass the entire scope of the resource.

The attempt to incorporate ecosystem principles into fishery management appears to complicate the matter of defining fishery management units. Ecosystem management needs to take place at a spatial scale that will encompass most of the key processes for ecosystem functioning. This line of thinking has led to the Large Marine Ecosystem (LME) concept and to the development of models at that scale (Sherman 1992; Sherman *et al.* 1993; Christensen and Pauly 1993). The scale of LMEs is such that 49 of them comprise the coastal areas of the entire world. Longhurst's (1998a) ecological geography of the sea takes the process of recognizing the large-scale spatial patterns in the oceans even further and provides the fishery manager with a basis for understanding the ecosystem context for fisheries management.

For most of us, conceiving of management at the spatial scale of the LME, and at a level of complexity that includes all ecosystem functions, is mind-boggling. It is difficult to find practical connections between these concepts and day-to-day management needs of small-scale fisheries. Yet we all know intuitively that these connections are potentially important. This lack of clarity is to be expected in an emerging area that researchers around the world are now grappling with. This should not deter the manager from attempting to bring these concepts to bear on the fisheries that they manage. However, for the manager who is attempting to define management units in practical terms, it may be useful to think primarily of management units that are defined in terms of the target and bycatch resources and the vessels and gears that exploit them. This definition should include ecological aspects such as essential habitat. Ecosystem considerations can be brought into the picture at another level, as the context within which the management unit functions. Thus, the key linkages between the management unit and the higher, ecosystem level can be considered explicitly as information becomes available, without allowing the lack of information to delay action at the lower level. Similarly, in management planning, measures can be included that aim to have an impact on the ecosystem that encompasses the fishery.

For practical purposes, the management unit should be defined to include the resources, fishers, and communities that have the strongest interconnections. There will always be an element of subjectivity in assessing what interconnections are sufficiently strong that the elements must be incorporated in the definition. There are no strict rules for achieving the appropriate balance between inclusion of interactions and the simplicity that is essential for management to be feasible. In this regard, stakeholder perceptions and acceptance could be strong guiding factors.

3.6 Fishery management objectives

The need for clear and clearly stated objectives has frequently been identified as necessary for the development of a workable fisheries management system (Pido 1995; Shepard 1991; World Bank *et al.* 1992). Often, the process of arriving at consensus regarding the objectives is an effective means of promoting an exchange of information and understanding among stakeholders.

3.6.1 The variety of benefits of fishery management

A wide variety of social and economic benefits may be derived from fisheries through management. Any of these benefits can be an objective of fishery management. The list in **Table 3.6** includes most of the objectives commonly stated for fisheries management. They appear to fall into three main groupings. The first relates to sustainability of the resource, ensuring that its productive capacity is assured into the foreseeable future (termed "biological" by Clark 1985). The other two groupings are economic and relate either to the optimization of returns from the fishery (efficiency) or to the distribution of those returns among stakeholders; that is, equity.

TABLE 3.6 SOME OBJECTIVES OF FISHERY MANAGEMENT.

OBJECTIVE	MAIN PURPOSE		
		ECONOMIC	
	SUSTAINABILITY	EFFICIENCY	EQUITY
1. Maximize catches		√	
2. Maximize profit		√	
3. Conserve fish stocks	√		
4. Stabilize stock levels	√		
5. Stabilize catch rates		√	
6. Maintain healthy ecosystem	√		
7. Provide employment			√
8. Increase fishers' incomes			√
9. Reduce conflicts among fisher groups or with nonfishery stakeholders			√
10. Protect sports fisheries		√	√
11. Improve quality of fish			
12. Prevent waste of fish	√	√	
13. Maintain low consumer prices			√
14. Increase cost-effectiveness		√	
15. Increase women's participation			√
16. Reserve resource for local fishers			√
17. Reduce overcapacity	√	√	
18. Exploit underutilized stocks	√	√	
19. Increase fish exports		√	
20. Improve foreign relations		√	√
21. Increase foreign exchange		√	
22. Provide government revenue		√	

Source: adapted from Clark 1985

Any of the benefits are valid objectives for a fishery, depending on the stakeholders' needs. However, it is not possible to achieve them all for a single fishery. Several of the objectives listed are incompatible within the same fishery, and there may even be effects between linked fisheries. Often, the different objectives relate to the interests of different stakeholder groups. For example, management would take different approaches in meeting the needs of recreational and commercial fishers, or of fishers and non-consumptive users of the resource. If the needs of the various stakeholder groups are not resolved at the stage of setting management objectives, inter-group conflicts arise during implementation.

It is not uncommon for policy or decision-makers to promise a long list of benefits, many of which are incompatible, that the fishery will provide under their direction. They then leave the fishery manager with the intractable problem of resolving the conflicts among users. Thus, the manager must educate and inform both the decision-makers and the users about what is possible and feasible for a given fishery, and must develop, with the stakeholders, a set of objectives that are internally consistent and acceptable (through compromise) to the stakeholders.

Broad statements of what fishery management will achieve are best used as mission statements, or, in the LFA context, as the goal of management. An example of a mission statement is "To develop manage the fisheries of Country X in order to obtain the optimum sustainable yield for the benefit of the people of Country X." However, in order to approach the management of a single fishery, it is necessary to focus on the objectives that are desirable for that fishery.

3.6.2 Discussing and communicating management objectives

The single-species surplus-production model and its bioeconomic adaptation (Gordon 1954) provide a useful basis for examining the relationship among several of the objectives mentioned above. These models, which have been a part of fisheries management for decades, are discussed in most texts on fisheries management and economics (for example, Christy and Scott 1965; Clark 1985; Hilborn and Walters 1992). For the purposes of a discussion on objectives, word and graphical versions of the models will suffice.

The biological component of the model is based on the observation that as fishing effort increases and the resource biomass declines, total yield increases to a maximum then decreases. The economic component of the model assumes that total revenue from the fishery will follow a similar pattern to the yield, although the shapes of the curves may differ if prices change with supply. It also assumes that the cost of a unit of fishing effort remains constant, so that the total cost of fishing is in proportion to total fishing effort (**Figure 3.3**). Therefore, the development of the fishery will reach a point at which fishing costs will exceed the revenue from fishing and the fishery will cease to attract new effort. This is the point toward which any unregulated fishery will tend (**Figure 3.3**). The lower the cost of fishing, the more depleted the stock will be at this point. Therefore, small-scale fisheries with inexpensive gear and fishers who are willing, or forced through poverty, to accept low returns, can result in severe depletion of fish stocks. This bio-economic adaptation has also been extended to multispecies fisheries by assuming that the entire assemblage of fishes will behave in a fashion similar to that of a single species.

Complexity and uncertainty in the quantitative application of these models may render them impractical for small-scale fisheries, or even large scale ones. This can be due to many factors, such as the way that the population responds to exploitation, technological changes in the fishery over time, the response of price to supply, or

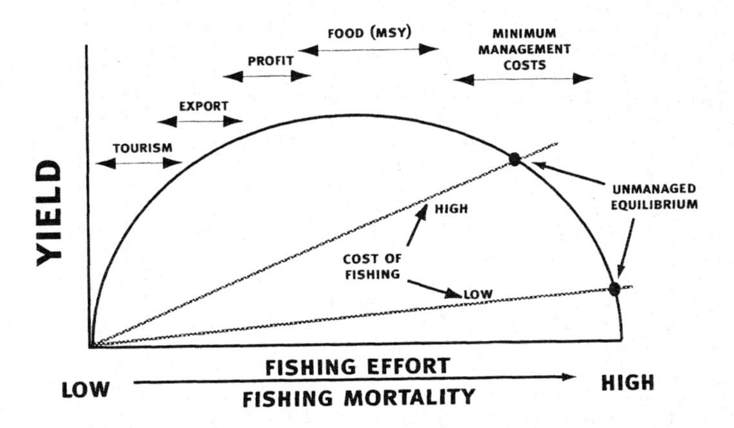

Figure 3.3 Some relative ranges for the levels of fishing intensity corresponding to different societal objectives for marine resource use in the context of a multispecies surplus production model for reef fish fisheries.
Source: Caddy and Mahon 1995

non-linearity in the cost of fishing. Furthermore, the application of these models may require sophisticated analytical techniques and large quantities of data. Nonetheless, the general form of the bio-economic models described above is sufficiently robust to be a reasonable basis for a discussion of management objectives.

Many of the objectives listed in **Table 3.5** can be identified with reference to this model (**Figure 3.3**). A fundamental question for fishery managers and policy-makers is "Where do we want to be on this curve?" Answering this question requires that the fishery manager and decision-maker develop an appreciation of the relative costs and benefits of being at a particular point on the curve. These issues are reviewed from right (most exploited) to left (least exploited) of the curve (**Figure 3.3**).

A fishery may be at its unregulated equilibrium by default, or due to the managers' conscious decision to not invest any money in managing the fishery. This decision may be taken for small fisheries that would be difficult to manage and are therefore perceived as not worth the investment. Unfortunately, the decision to leave the fishery to reach its unregulated equilibrium is sometimes rationalized by the erroneous view that because this state of the fishery corresponds to the point of maximum effort, allowing it to remain there maximizes employment, which is then cited as the (*ad hoc*) management objective.

The view that employment of fishers is maximized at the unregulated equilibrium is sometimes held because at this stage in small-scale, artisanal fisheries, there is usually a high number of individuals engaged in fishing. However, given the low catch rates, the average return from the fishery to each of these individuals will be less than if the same number of individuals each exerted less fishing effort, so that the total effort was that which corresponded to maximum sustainable yield (MSY).

At that level of total effort, the same number of fishers could be supported, each taking a larger catch than is possible at the unregulated equilibrium. Alternatively, the increased total catch of MSY could actually support an increase in the number of fishers over that which would be supported at the unregulated equilibrium. Therefore, the objective of maximizing the employment of fishers is likely to coincide most closely with the objective of maximizing the catch.

If the employment objective applies more broadly to total employment in the fishing industry, then there is even more reason to take the view that maximizing catch will maximize employment, because there will be more fish for processing and distribution operations.

When the objective is to maximize the production of fish for food, maximum sustainable yield is frequently identified as the target reference point. However, due to imprecision in determining MSY, and the natural unpredictability to most fish stocks, MSY should be perceived as a limit that should not be exceeded rather than as a target (Caddy and Mahon 1995). Managers will often seek to leave a safety margin, and two-thirds MSY is often recommended as a target reference point when the objective is to maximize the production of fish for food.

When profitability of the enterprises engaged in fishing is the primary concern of the managers, the target will be the maximum economic yield (MEY). The location of MEY varies depending on the slope of the cost line. As cost-per-unit-effort decreases, MEY approaches MSY and may, in low-cost fishing operations, exceed two-thirds MSY.

In a multispecies fishery, such as for reef fishes, the proportion of large, economically preferred species required for export may become unacceptably low at levels of fishing effort that are lower than those that would produce MEY (in terms of local currency) or two-thirds MSY. If maximum foreign exchange is a high national priority, it may be necessary to impose more rigorous controls than would be required for the two above objectives. For example, in a reef fish fishery it may be necessary to restrict fishing to hook-and-line gear.

In addition to the extractive benefits that may be derived from fisheries, non-extractive benefits may assume a high priority in countries with a well-developed tourism industry. If a fishery harvests the same fish resources that draws tourists for snorkelling, diving, glass-bottom-boat viewing and/or sport fishing, and tourism is a high national priority, then maintaining optimal fish assemblages for tourism may be a management objective. Because these uses are generally incompatible with intensive fishing, direct conflicts between users may occur. In some cases, non-consumptive uses may be given higher priority than fishing. For example, the reef fishery was closed in Bermuda in 1990, when tourism was given higher priority than fishing (Butler 1993).

3.6.3 Conflicts among objectives

Any two objectives that require the fishery to be at different points on the curve will be in conflict. The manager must find a way to decide among the conflicting objectives. Where equity issues are involved, the manager must either have the decision-makers rule in favour of one group or, preferably, work with stakeholders to reach a compromise. There are methods for working with stakeholders to reach a consensus on management objectives. One of these is the Delphi technique, an iterative process in which stakeholders individually provide information on their view of a situation. The information is then fed back to the group in a format that ensures anonymity of the individual. The process is repeated until the changes in overall group perception are relatively small. In many cases, the stakeholders are influenced by one another and converge toward a consensus.

Another, more sophisticated, approach is multiattribute or decision analysis. Healy (1984) remarks that "The failure of a multiattribute analytic methodology for dealing with problems like OY [optimum yield] to emerge in fisheries is surprising, since a well-developed methodology has been effectively employed in other areas of natural resource management to incorporate social, economic, and ecological goals into a single analysis." He provides an introduction to the rich literature on multiattribute analytic techniques and an overview of the process (**Table 3.7**), with examples for Skeena River salmon and New England herring. Shea (1998) points out that the value in a decision theoretic approach may be as much in using it as a framework as in the complex mathematical details of the analysis. The framework shown in **Table 3.6** can be used with intuition, rules of thumb and experience to produce robust and useful decisions.

Table 3.7 The six steps in the application of multiattribute utility theory to determining optimum yield.

1. **Bounding the problem:** Deciding which factors and which constituencies are to be taken into account and which are to be left out in determining optimum yield
2. **Determining the feasibly policy alternatives:** Deciding what is the range of technically feasible yields within which the optimum yield must lie
3. **Deciding on the attributes of the problem:** Selecting a comprehensive, mutually exclusive, and preferentially independent set of attributes
4. **Setting the attributes weights:** Determining the relative importance of attributes as criteria for distinguishing among yields
5. **Scoring the policies:** Objectively scoring each feasible yield against each attribute
6. **Applying the decision rule:** Combining the scores and attribute weights for each yield according to the predetermined decision rule and selecting the optimum yield

Source: from Healey 1984

As Healy points out, the decision analysis approach has not often been applied to fisheries. Two recent examples are the Campeche Bank octopus fishery (Diaz-de-Leon and Seijo 1992) and the Chilean loco (Tam *et al.* 1996). These illustrate the use of systematic methodology in resolving complex inputs to the setting of fisheries objectives.

The reduction or resolution of conflicts among objectives is a complex matter that is part of many of the equity issues. An example of an area of conflict could be the use of a resource by fishers considered to be "outsiders" (Derman and Ferguson 1995). Outsiders with more financial backing, mobility, and sophisticated gear may out-compete local fishers with traditional gear, causing social and economic hardship. In addition to the methods described above, there are formal processes for mediation and conflict resolution that can be used to address differences in objectives among stakeholders (for example, Fisher *et al.* 1991; Bush and Folger 1994; Fisher and Ertel 1995)

3.6.4 Prioritization of objectives

Fishery management objectives can be prioritized within the same fishery to construct a hierarchy. Generally, the selected location of the fishery on the production curve will be the primary objective, and will determine the main thrust of fishery management. However, subject to constraints due to the primary objective, there can be secondary management objectives.

For example, maximizing total employment could be a primary objective, which would probably be achieved by fishing as close to the top of the curve as can be sustainably achieved (which may not be very close at all). However, management would then have to become involved in how that catch was taken; for example, by deciding between operations that employ much technology and few fishers or those that employ little technology and many fishers. Within either of these two options, employment can be increased by distributing the catch among the largest possible number of fishers while still allowing each to catch enough fish to remain viable. This requires access control.

If maximization of foreign exchange were the primary objective, one could still have a secondary objective of increasing employment or efficiency in the way fish are harvested, or in the quality of fish landed, etc. With that secondary objective of maximizing employment, one would need measures similar to those used when it is a primary objective, but within the constraints of the primary objective.

As the example above illustrates, objectives that would conflict if both are primary may be accommodated if one is primary and the other secondary. Subconscious awareness of this may lead managers and policymaker s to state conflicting objectives for a fishery. When the hierarchy is not explicit, the various stakeholders may each assume that their preference has priority. Therefore, one must prioritize the objectives and state these priorities clearly. This transparency is necessary so that stakeholders may contribute to the meeting of objectives.

3.6.5 Application of the precautionary approach in setting objectives

One way of seeing the sustainability of benefits is as constraints or limits within which the other objectives may be achieved. In that way, the sustainability of the resource becomes the primary objective. However, there is no absolute indicator of the level of fishing at which the resource becomes unsustainable. Owing to natural variability and other sources of uncertainty, the "stock assessment refiners" have recently been devoting much of their effort to estimating the risk of becoming unsustainable at various levels of fishing. There are also different levels of risk to be considered:

- The risk of the fishery entering an undesirable state;
- The risk of the ecosystem suffering long-term disruption;
- The risk of extinction of the exploited species.

The level of risk acceptable in the latter category has been determined by international agreement (*Convention on Biological Diversity*). However, the levels of risk acceptable in the first two categories are not well defined and are largely a matter for stakeholders to determine. This is an area in which there may be disagreement among large-scale fishing, small-scale fishing and public interests in the fishery. The former group may discount the future value of the fishery or ecosystem, preferring to have the benefits now, in the form of large catches. The latter two will take a longer-term perspective on the future value of the resource.

Even if the long-term view prevails in objective setting, there will be difficulty in agreeing upon an appropriate level of risk. The precautionary approach is about erring on the side of conservation of the resource: "When you don't know, don't go." The science required to know is costly; therefore, precaution will inevitably be a major component of the management of small stocks.

The setting of objectives for small stocks should include some statement of what would constitute unacceptable biological circumstances. An objective would therefore be to avoid these circumstances, thus setting limits for the fishery. Determining what to do, or what not to do, to stay within limits would be a part of the process of assessing the fishery. Clearly, this will vary from one resource type to another, depending on the biological characteristics of the resource. Longhurst (1998b) points out that owing to the life history differences between yellowfin tuna and Atlantic cod, the former will probably continue to provide substantial yields with minimal management, whereas the latter cannot. Tropical shrimp stocks are also a resource type at minimal risk of being overfished to the point of collapse.

3.6.6 Nonfishery impacts on fishery management objectives

Fishery management systems will be frustrated by the effects of entities such as trade agreements and conservation thrusts that are peripheral to fisheries. Consequently, the setting of realistic objectives requires an understanding of the linkages between fisheries and these entities. Because conservation groups are becoming key players, it is increasingly important to perceive them as stakeholders and to include them in the planning process from the earliest stage. As discussed in relation to **Figure 3.3**, other sectors of the economy, particularly those that depend on the same coastal and marine areas as do fisheries, may have needs or make demands that override fishery concerns. For example, foreign policy in marine affairs and trade can have significant effects. These factors need to be assessed and taken into consideration in objective setting.

3.7 Transparency: Documentation, Communication, and Participation

In the new approach to management, transparency is important if the management system is to be a learning one. This is particularly important for a process with uncertain outcomes. Therefore, documentation of process, plan, and outputs of and inputs to process are necessary for the system to work. Transparency is also important if stakeholders are to be properly informed so they can participate fully. For some stakeholder groups, the process must go further, educating them to prepare them to participate.

3.7.1 Documentation

To be truly available, the best available information must be in a form that can be shared with stakeholders. Therefore, the practice of documenting available information in a form that makes it available should be given emphasis in fishery management. There are two aspects to this: documenting information and managing the information. Focusing first on documentation, most managers and their staff are continuously acquiring information: through literature review, research projects, field observation, and interaction with stakeholders. To make the most efficient use of the limited staffing that is typical of management units for small-scale fisheries, this information should be documented in useable units that are kept within a framework that allows easy access and retrieval.

This book has already emphasized the importance of local and traditional knowledge. However, this knowledge is seldom systematically compiled and documented. A system to capture this type of information can be as simple as a filing cabinet with file folders for various categories of information. But someone must have the responsibility for encouraging the collection of the data, and for synthesizing it from time to time. The acquisition does not need to be a large project. The information can be acquired gradually as personnel go about their duties in the field. Once the system for capturing it is in place, the information will grow over time into a useful database. This may also serve as a useful backup to more technological systems.

The case of landing sites in Jamaica provides an example of how an ongoing system to acquire data from fisheries department staff can be useful for management. There are about 250 landing sites around Jamaica's 600 km of coastline. Many of them are officially designated, but there are numerous unofficial ones as well. The

Table 3.8 A form that could be used to document landing-site information in Jamaica.

LANDING SITE DATA FORM

Landing site: *Barmouth*				**Beach policy #:** 104F
Parish: *Clarendon*	**Other names:** *Portland Cottage, Portland Beach, Bournmouth*			**Fisheries #:** 102F
Vessels and gear				
Type/description	**Number**	**Prop.**	**Crew**	**Type of activity**
Wood canoe	9	Un-mech	2	*On shelf, Chinese net, pots*
Wood canoe	2	Mech		
Fibreglass canoe (reg.)	36	Mech	2-3	*Pots, Chinese nets and lines, many on Pedro Bank*
Fibreglass canoe (large)	5	Mech	2-3	*Have ice boxes, go to Pedro Bank*
Total	52	*Fisheries Division 1998 estimate = 34*		

No. of fishers; Licensed = *119* **Estimated** = *130*

Boat ownership: *Fishers' own vessels*

Description of fisheries: *Usual mix of gears and fisheries. About 20+ spear fishers operate from this area regularly; some occasionally come from as far away as the north coast. Many of the larger vessels go to Pedro Bank.*

Where fishers live: *Immediate area*

Where fish sold: *Vendors and immediate area*

Storage sheds x	Toilet	Water x	Fuel x	Instructor	Ice	Lighting
Electricity x	Coop/ Assoc	Coop Office		Artisans x	Vendor stalls	Parking

Description of landing site: *Short road off the Portland Point Rd. about 3 miles south of the turning to Rocky Point. The road goes between houses, down to a creek through the mangroves. The boats are moored in the creek, and some are pulled up on the mud apron by the creek. The creek enters into West Harbour.*
There are no facilities.

Problems: *Lack of facilities.*

Notes: *CCAMF formed a fisher association on this beach and is planning to acquire a lease for the fishers.*

Previous surveys/information documents that included this site: from 1963, 1981, 1995, 1998 situation is dynamic, with landings sites changing as users move between them, add structures, and so on. Over time, landing sites may shift location in response to markets and coastal development, or may just change in importance. Jamaica has historically carried out a fishery survey once every 10 years. A lack of funds extended the last interval to 18 years. Thus, toward the end of an interval, the situation may be very different from the one described by the previous survey.

Fisheries department staff at the headquarters in Kingston do not have the resources to update the survey information at shorter intervals than the main survey. However, since they visit landing sites in the course of their duties, they see the changes that have taken place or could check for changes with little additional effort. In addition, fisheries field personnel are knowledgeable about the conditions in their regions. The problem is that there is no means in place to capture information from the field or headquarters staff after they have visited the field. A simple data capture system, such as **Table 3.8**, with a separate sheet for each landing site, would provide the means for personnel to compare their observations to the information, record changes and have them entered into a master record. Similar systems could be put in place to record a variety of information acquired from fishers in the course of interacting with them. Formally or informally publishing the accumulated information from time to time is one means of ensuring that it also becomes more accessible.

3.7.2 Participation

Several methods can be used to promote participation in management planning. Navia and Landivar (1997, Section VII: Methodologies, Approaches and Techniques for Participation, www.iadb.org under policy) review some methodologies. Two other excellent sources of information on participatory methods are the *International Institute for Rural Reconstruction Manuals on Participatory Methods in Community-based Coastal Resource Management* (IIRR 1998) and the *Change Handbook* (Holman and Devane 1999). Technology of Participation (ToP), developed by the Institute of Cultural Affairs (ICA), one such methodology, provides a suite of methods that can be used with fishery stakeholders, especially for planning. The methods, briefly described as an example in **Box 3.1**, are highly participatory, visual and do not require a high degree of literacy among participants. They can be self-taught or, preferably, the user can take part in the short courses offered by ICA ToP trainers. These can be arranged in country and will interest persons from both public and private sector, which may make the course affordable. As described above, the use of methods to deal with multi-user situations can be taken further with the use of multiattribute objective analysis and similar methods.

Transparency and stakeholder participation can play pivotal roles in gaining the support of decision-makers.

Box 3.1 Technology of Participation (ToP) methods.

Technology of Participation, a group-process methodology developed by the Institute of Cultural Affairs (ICA)(Spencer 1989), has a long history of use in community development (Navia and Landivar 1997). It is designed to help groups think, talk and work together.

ToP is a set of processes — Action Planning, Strategic Planning and so on — comprising basic methods that are adapted to the purpose of the process. The basic methods, as shown in this table, include individual and group brainstorming, the workshop method that clusters the group's ideas into categories, and focused conversation for group reflection.

OVERVIEW OF TECHNOLOGY OF PARTICIPATION METHODS AND PROCESSES

Processes and methods	Outputs
Basic methods	
Workshop method (WS)	Consensus
Focused conversation method (FC)	Conclusion and decision
Strategic planning process	
Find the vision (WS/FC)	Vision
Identify the blocks (WS/FC)	Blocks
Develop strategic directions (WS/FC)	Start Dir
Prepare action plan	Action plan
One-year calendar	Roles and responsibilities
90-day implementation calendar	Implementation plan
Priority wedge	Priorities
Reflect on process (FC)	Conclusion and group decision
Visioning process	Vision and statement
Vision workshop (WS)	Vision elements
Reflection (FC)	Conclusion and decision
Action planning process	Action Plan
Victory circle	Short-term vision
Current reality	Team strengths and weaknesses
Commitment	Team commitment
Key actions (WS/FC)	Key actions
One-year calendar	Roles and responsibilities
90-day implementation calendar	Implementation plan
Reflection (FC)	Conclusion and group decision

Basic methods

The ToP workshop method brings out a group's ideas and focuses their attention on the issues. This design and problem-solving tool, which generates creativity and consensus by bringing all participants into the process, consists of five steps:

Context >>> Brainstorm >>> Order >>> Name >>> Reflection

The purpose of the workshop is determined beforehand, usually in the form of a focus question, which is the context. Participants begin by generating ideas individually and then discussing them in small subgroups (three or four individuals). They select their subgroup's most relevant ideas, which are then clustered by the overall group on a wall. The group gives each cluster a name that captures its concept. This information then becomes input into planning.

ToP-focused conversation is a group-discussion method in which the discussion leader asks questions designed to bring out responses on four levels:

1. Objective, to bring out relevant data (based on objective facts);
2. Reflective, to bring out feelings, memories and associations;
3. Interpretive, to inquire about significance and meaning; and
4. Decisional, to move the group toward decision and action.

Thus, the group moves by stages from information to decision. (This method is also referred to as ORID.)

Participatory strategic planning process

This two-day planning process includes four workshops using a sticky wall and four variations of the ToP workshop method. They are:

1. Vision
2. Underlying contradictions
3. Strategic directions
4. Action planning.

The process starts with the creation of a shared vision that defines where the group wants to be, with respect to the focus question, in three to five years. After identifying their shared vision, the group takes part in a workshop to name the blocks, or contradictions, to the vision. Next, the group develops strategies to remove the blocks and completes the process by creating an action plan for implementing the strategies that will allow the group to move toward its vision.

Action planning

The ToP action planning process allows stakeholders to define their success before looking at the problems associated with a project and determining their commitment based on their strengths and weaknesses. This method is used to plan short-term projects or campaigns and for prioritizing activities over a period of time (Spencer 1989). There are five steps to the process:

1. Name the victory, to describe what the success will look like;
2. Assess the current reality, to determine the group's strengths and weaknesses and the likely benefits and dangers of success;
3. Name the commitment, which is what the group is able to do based on its strengths and weaknesses;
4. Determine the actions required to achieve the victory; and
5. Create a calendar that includes a timeline of actions with the names of the people who will carry them out.

The actions, calendar and assignments are prepared using the workshop method.

Box 3.2 provides an example of how Technology of Participation methods were used to get fishers to participate in developing an approach to managing a fishery.

Box 3.2 Sea Urchin Comanagement Project: Meetings to determine the future of the sea egg fishery.

WHAT WAS DONE

After small group discussions with more than 100 sea egg fishers around Barbados, 37 fisherfolk representatives from 17 communities used the ToP strategic planning process (**Box 3.1**) to determine the future of the depleted sea egg fishery. Over the course of three meetings, they worked together through the four stages described below.

STAGE 1. THEY DEVELOPED A VISION BY ASKING, " WHAT DO WE WANT TO SEE IN PLACE IN FIVE YEARS FOR THE SEA EGG FISHERY?"

- Sea eggs back and divers working
- Management measures for the sea egg fishery decided upon and in place, including licensing
- Sea egg divers' organization established
- Fisheries and sea egg divers working together, comanaging the fishery
- Safer harvesting
- Marketing system set up for sea eggs
- Laws against poachers more strictly enforced
- Having some effect on pollution and polluters
- Research and development activities ongoing.

Stage 2. Then they asked, "What is keeping us from getting there?"

- Divers don't cooperate with one another
- Government and fishers don't communicate with each other
- There are wrong and inadequate rules and regulations and inadequate law enforcement
- Government is not dealing with polluters and puts a low value on fisheries.

Stage 3. Then they developed some strategic directions by asking, "What will move us ahead?"

Strategic Direction # 1: Cooperating for the betterment of the industry

- Divers form an organization
- Get meeting attendance from others
- Bring experienced fishers into Fisheries
- Formulate a comanagement plan.

Strategic Direction # 2: Teaching people the value of the sea egg fishery

- Teach people about conservation
- Promote the value of sea eggs by lobbying.

Strategic Direction # 3: Improving laws and enforcement

- Develop methods for making enforcement work
- Protest against pollution.

Strategic Direction # 4: Trying new methods

- Restocking the sea eggs, either from abroad or from another site on the island.

Stage 4. Then they made a plan of action for the first year.

In the first quarter we will:

- Do an outreach program to interest other divers, giving them information about what we have done and getting them involved.

In the second quarter we will:

- Distribute a brochure with information about sea eggs and the sea egg fishery.
- Begin to lobby the Department of Agriculture and the Department of Environment with petitions for action.

In the third quarter we will:

- Hold the first meeting of the law-enforcement committee
- Hold a general election for officers of the Association
- Request information from the government about pollution
- Consider whether we want to try restocking sea eggs.

In the fourth quarter we will:

- Hold the first comanagement meeting with the Association officers and core group
- Make recommendations for ways to manage the sea egg fishery.

The results of the process shown above are relatively simple, but it gave the participants their first opportunity to share their views and concerns and to realize that these were held in common with their fellow stakeholders. The plans that were made are also simple, common-sense measures. If these are followed, there will probably be no need for an assessment of yield from the fishery. Instead, the recovery of the resource can be monitored and adaptive measures adopted through an extension of the above process.

These meetings were also an opportunity to report on information that was gathered from fishers. This included traditional ways that fishers managed the sea egg fishery. When sea eggs were abundant, fishers used to: check a few sea eggs from a patch to see if they were ripe, and, if not, they would take no more sea eggs from that patch; take only the ripest sea eggs from the edge of a patch and leave those in the center to ripen; leave large individuals scattered throughout the fishing area as breeders; drop small unripe urchins in places that had been fished out; and break sea eggs on the beach and bury the husks so that they were not disposed of on fishing grounds. Fishers also shared their views of the many changes that had happened over the years in the ways that sea eggs are harvested, and acknowledged that some of these changes had decreased the abundance of the crop. This shared information provided a common base from which to move forward.

Much of fisheries management's failure to conserve resources has been blamed on a lack of political concern or will. A lack of political concern may stem from the common perception that fisherfolk lack power and are therefore of little importance in the political arena. This situation can be improved by any measures that empower fisherfolk. When there is participation and consensus, fisherfolk representatives are able to speak with the knowledge that they are supported by a constituency. Politicians understand this type of situation and will respond favourably.

Even when they *are* empowered, fisherfolk not participating in management, and thus not informed about what the fishery needs and in agreement with the plans to achieve it, may use their power to have decisions made that are in *their* short-term interest but bad for the fishery in the long term. Faced with conflict between demands from fisherfolk and technical advice from managers, politicians often give in to the short-term needs of voters. The negative consequences of unwise management decisions are frequently outside the time frame of political terms of office. Fisherfolk's participation in formulation of management, seeking consensus wherever possible, can go some way toward reducing the frequency with which decision-makers face conflicting demands, but this does not eliminate the need to deal with conflicts.

3.8 Conclusion

This chapter expanded on approaches to planning processes for managing small-scale fisheries to meet agreed-upon objectives. This approach is based in the reality that for most small-scale fisheries, conventional assessment and management approaches may not be justifiable on the basis of the total value of the resource. Thus, key elements are stakeholder participation, a precautionary approach, and the use of a wider variety of information types and sources than may be typical for conventional fisheries assessment.

In view of the above elements, we have emphasized proper planning and design of the planning process. Process is essential to participation and transparency, and clearly stated and agreed objectives are the foundation of planning.

Chapter 4
Fishery Information

4.1 Introduction

This chapter examines the information acquisition, management and use needs of small-scale fisheries. We have already emphasized that management of small-scale fisheries will require the use of a wider variety of information types than has been the practice in conventional fisheries management. In part, this reflects a shift in focus from assessment of the resource to assessment of the fishery as a whole. It also reflects a trend toward the incorporation of common, or stakeholder, knowledge (including traditional knowledge) in management systems. The acquisition, management, and use of this type of information is not readily accommodated by conventional, numerically based data and information systems, which are presently heavily biased toward biological and economic parameters. The alternative, more holistic, direction presents challenges in the collection and management of information. But if these challenges are overcome, improved fisheries management should result.

For small-scale fisheries, limited resources dictate that information gathering and management be efficient. Therefore, we emphasize the need for process and focus. Information can be a costly commodity, and more is not necessarily better. If funds are to be used effectively, information must be acquired and managed in the context of a plan with a clear view of how it will be used in management. Since time also costs money, and information requests can be urgent, we have emphasized the need to develop rapid appraisal techniques for all aspects of small-scale fisheries. Some of these techniques are already available, others are under development, and yet others need research attention.

This chapter is more oriented toward principles and directions than methods, to provide guidance about where a manager might want to invest effort. For details on methods that might be useful, we will direct the reader to the appropriate manuals and also to Chapter 5 of this book, which provides considerable detail on frameworks and methods for gathering information in ways, and from sources, that differ from those used in conventional fishery assessment.

4.2 Data and information collection

Information is the product of data that have been acquired, analyzed, and interpreted for use. The distinction between data and information is not important for this discussion. However, it draws attention to the need for vigilance over the extent of filtering and manipulation to which data may have been subjected by the providers and collectors before they become accessible to the fishery manager. Many rapid appraisal methods make use of data that have already been informally interpreted (Chambers 1997), and these should be used with caution.

4.2.1 Rapid appraisal

Rapid appraisal is an emerging methodology of considerable interest, but less well known to fishery scientists and managers than most conventional methods. The term has been used to describe approaches to information gathering that provide alternatives to the conventional sampling and census-based methodologies that dominate scientific research. Rapid appraisal techniques allow the quick acquisition of key information that is perceived as essential to management decision-making. Pido *et al.* (1996) provide a brief historical and conceptual background to rapid appraisal, as well as a manual for the rapid appraisal of fishery management systems. A variety of rapid appraisal techniques are also provided in the manual by IIRR (1998).

One cannot draw a precise dividing line between conventional research methods and rapid appraisal methods, as the latter may often be adapted from the former. Rapid appraisal methods may range from interviews with key informants to scaled-down versions of conventional sampling, several examples of which are dealt with in the next chapter. Rapid appraisal techniques can be used for acquiring a wide variety of types of information: social, economic, institutional, organizational, technological, and biological. Thus far, however, the emphasis in rapid appraisal methodology has been on the human aspects of fisheries (Pido *et al.* 1996).

4.2.2 Appraising fishery resources

Rapid appraisal techniques for the resource and its environment have been oriented mainly toward habitats and environmental impacts. Techniques that would provide indicators of the status of fishery resources have received much less attention. There has been some work on the application of conventional fishery data collection methods and stock assessment methodology to small-scale fisheries, but it cannot really be termed rapid appraisal. In particular, Caddy and Bazigos (1985) provide a variety of methods for acquiring fishery information in manpower-limited systems. To a large extent, the approaches that they describe require participation of industry stakeholders, particularly fishers, in providing information either directly to officials or by filling out forms and logs. Similarly, an approach to fishery catch and effort data collection based on path diagrams aims to maximize the use of systems that are already in place while ensuring that all pathways are covered by the simplest available methods appropriate to the pathway (Mahon 1991) (**Box 4.1**). Although aiming to make data gathering simpler and more efficient, neither of these initiatives qualifies as rapid appraisal. Nevertheless, near-rapid or simplified methods deserve some attention, since they bridge the gap between the conventional and the alternatives.

Box 4.1 Fishery pathways and data collection.

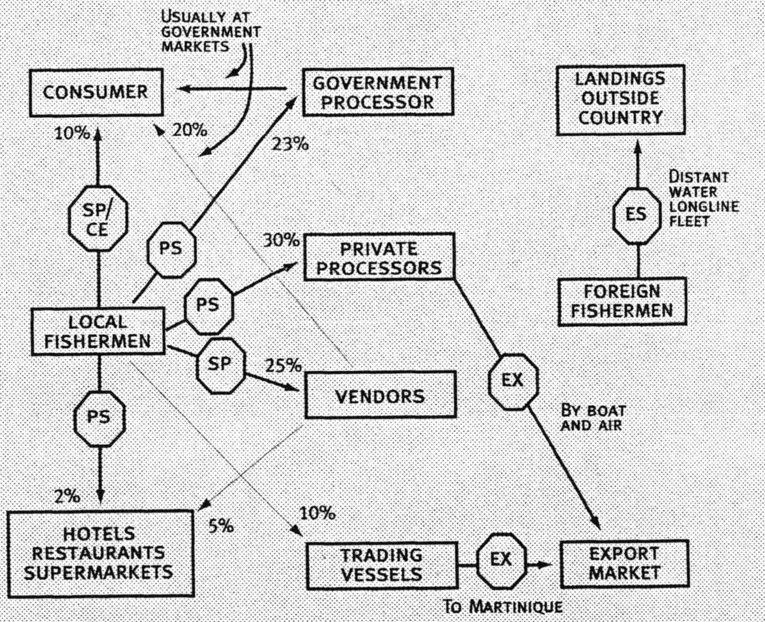

Path diagrams are a useful means of summarizing and communicating the interactions among stakeholders in a fishery. This example shows a relatively complex system for large pelagic and "quality" hand line-caught demersal fishes in an eastern Caribbean island. For some fisheries, such as trap fisheries for reef fishes or small coastal pelagics, the path may simply be fisher to consumer. Percentages indicate the amounts of product moving along each pathway. The pathways also provide the means to identify suitable points in the system for applying various data collection methods (in hexagons). Data collection instruments shown below include: sampling (SP); census (CE), when numbers are few; purchase slips (PS) when businesses with accounting systems are involved; and export data (EX) when government export permits are required. For foreign-caught fishes, the only approach may be to estimate (ES) from data provided by international organizations such as ICCAT (adapted from Mahon 1991).

Much of the early work on assessment of small-scale stocks was aimed at tropical stocks to determine if they behave in the same ways as the temperate stocks for which stock assessment methods were mainly developed. The conclusions have generally been that the methods can be applied, in spite of constraints such as difficulty in determining ages, ill-defined recruitment periods due to multiple or continuous spawning, and the greater prevalence of multispecies fisheries (Munro 1979; Appeldoorn 1996). An aspect of this work has been the issue of whether the dynamics of aggregates of species in multispecies fisheries is similar to that of single species, and can therefore be assessed collectively with single-species methods. Another focus of assessment on stocks exploited by small-scale fisheries has been the use of simplified methods or short cuts derived from more complex conventional stock assessment methods (Pauly 1979, 1983). However, most of the effort expended toward improving capability for assessing tropical small-scale fisheries has been oriented toward making conventional, single-species methods more widely accessible and user-friendly (Sparre and Venema 1992).

There is a definite need for more research on rapid appraisal indicators in general, and in particular for those reflecting the status of fishery resources. Some examples in Chapter 6 include species composition of the catch from a multispecies commercial landing and average fish length and percentage of mature individuals in the catch in relation to estimated maximum size.

4.2.3 A new multidisciplinary approach

A recent initiative regarding rapid appraisal of fisheries using a variety of characteristics is "RAPFISH" (Pitcher *et al.* 1998; Pitcher and Preikshot 2000). This method uses biological, technological, social, economic, and "ethical" characteristics to compare fisheries. Ethical characteristics include occupational alternatives, equity in entry into the fishery, and just management. In addition, a set of variables indicates the extent to which the fishery is in accord with the *Code of Conduct for Responsible Fisheries* (Pitcher 1999). Several fisheries can be compared, or a single fishery can be examined over time with regard to a subset of variables from any of the above categories, or all can be examined together. Fisheries are located in a two-dimensional space between the extremes of those that are considered "good," having all desirable characteristics, to those that are considered "bad," having all undesirable characteristics. The position of a fishery in this space is an indicator of its status and what is right or wrong with it. Most of the data used in this analysis is easy to acquire; variables are scored on a scale from 1 to 5, with intermediate values being acceptable. The analysis itself is sophisticated and requires knowledge of multivariate analysis. Nevertheless, the approach has the potential to be used as a diagnostic tool for small-scale fisheries management.

4.2.4 Expert judgment and experienced appraisal

Although actual measurement is still preferred in most situations, managers should be prepared to rely on the judgment of experts where appropriate or where there are no alternatives. Given errors in measurement and the costs of conventional data collection, expert judgment may be both equally useful and more affordable when compared with the usual methods.

Whereas expert systems can be quite formal (Mackinson and Nottestad 1998), expert judgment can be informal but still extremely useful. For example, a fisheries officer, researcher or fisherman of many years' experience and observation of different fisheries and states of exploitation is a repository of a wealth of knowledge that can be applied at a glance. Without the assistance of conventional data collection, he or she can visit a fish landing site or snorkel over a reef and quickly assess what is wrong or how well managed the fishery is. Common sense suggests that this ability is widespread and should be used where available, at least to formulate indicators, assess trends and determine areas that require more systematic information gathering. How we incorporate these observations into the data collection scheme is situation-specific, depending in part on the availability of other data for corroboration and the extent to which experienced observers agree with each other.

4.3 Traditional ecological knowledge

Preliminary fishery assessment and qualitative indicators of change can often be based on readily available fishers' knowledge of the catch trends, their observations of ecology and fish behaviour, and other information resulting from years of practical experience. Sometimes these aspects of knowledge may be handed down for generations, becoming the traditional ecological knowledge introduced in Chapter 2.

4.3.1 Applicable research methods

This section introduces the characteristics and uses of methods of collecting traditional knowledge. The first three (seasonal calendars, participatory mapping, and transects) summarize widely used techniques adapted from Rapid Rural Appraisal and Participatory Rural Appraisal (Chambers 1997; Grenier 1998). The next five (participant observation, semidirected interviews, key informants and focus groups, local and oral histories, and short questionnaires) summarize ethnographic research methods that have proven useful in traditional knowledge research.

4.3.1.1 Seasonal calendars

Groups of fishers summarize their major species and seasonal catch patterns. The information may be entered on a circular calendar, with each month represented as a pie in the circle and each species represented by a continuous or broken circle. A large-size circular calendar, which can accommodate information on 10 to 15 major species in an area, provides a convenient snapshot of the seasonal fishing pattern. The method helps the researcher understand the activities of fishers, but care must be taken to make the calendar from the fishers' point of view. For example, in Zanzibar, Berkes and colleagues found that the local Swahili months (lunar calendar), rather than the Christian calendar, were appropriate because fishing for species such as octopus strictly followed the tidal cycle.

4.3.1.2 Participatory mapping

The researcher provides a map of appropriate scale on which groups of fishers mark their major fishing areas as ellipses or polygons, by species. Transparent overlays may be used to create layers of maps by species, by season, and by fishing community. If collected systematically, the various layers can be entered into a Geographic Information System (GIS). Sufficient notice should be given to the community so that the major fishers are present at the meeting. Typically, fishers self-organize and one fisher marks the map while others provide additional detail and corrections. Such maps help the researcher figure out the relative importance of features such as reefs or fishing banks, whether or not there are community-based territories, and the major activity areas of communities of fishers in relation to MPAs, as was done, for example, in the Misali Island Conservation area, Zanzibar, Tanzania (Abdullah *et al.* 2000).

4.3.1.3 Transects

The researcher and key informants conduct walking tours (transect walks) through the coastal zone used by a community of fishers, or a boat tour across an area, to observe, to listen, and to identify different resource areas, as well as to ask questions to identify problems and possible solutions. Carried out by skin divers pulled by boats along a predetermined grid, this method can be adapted to conduct rapid surveys of coral reefs and seagrass beds (Pido *et al.* 1996; English *et al.* 1997). The various uses of the transect method provide an effective way for an outsider to learn about a local area, its features and its use, and for the local people to explain their point of view to a researcher or manager.

4.3.1.4 Participant observation

The researcher takes part in the activities of the fishers to learn by direct observation and experience. Widely used by anthropologists but time-consuming, this is the single most effective technique for understanding and appreciating fishing practices, social organization and informal rules. Especially important for the manager, participant observation reveals if there are institutions for the management of the commons (Berkes 1999). It also provides the insider view on resource abundance/crises, enforcement problems/solutions, and, in general, how the fishers make their livelihood (Jorgensen 1989).

4.3.1.5 Semistructured interviews

Semistructured or semidirective interviews provide an informal, flexible listening technique with open-ended questions, such as "Can you tell me more about the use of this fishing bank?" or "When do the tunas arrive and what direction do they come from?" (Huntington 1998). Using a checklist of topics instead of questionnaires, semi-structured interviews allow for more depth than do standard interviews. New topics and potentially interesting questions become apparent as the interview develops, and the informant plays an active role in guiding the interview. The technique also works well with group interviews, often supplemented by participant observation.

4.3.1.6 Key informants and focus groups

In traditional knowledge research, finding the "right" people is very important. The most accessible persons in a community are often not the best informants. Rather, the researcher should seek informants who are regarded as experts by the community or by their peer group of fishers. The research may be carried out with one (or a small number of) knowledgable and respected individual(s), as done by Johannes (1981) in his extremely detailed studies in Palau, Micronesia. A group of key informants brought together to brainstorm on a specific topic or issue is called a focus group. For example, focus groups were used to manage conflict and develop comanagement arrangements in the Cahuita National Park, on the Atlantic coast of Costa Rica (Weitzner and Fonseca Borras 1999).

4.3.1.7 Local and oral histories

Every fishery and fishing community has a historical context that is important in understanding why a group of fishers behave as they do (Baines and Hviding 1992). For example, finding out if a fisher's father and grandfathers were also fishers helps establish the likely depth of that person's knowledge about the fishery, given that traditional knowledge is multigenerational and cumulative. Historical information can provide an account of how things have changed or are changing. For example, histories can be developed to trace cycles of resource crises and management solutions over a multidecade scale, as done for Ibiriquera Lagoon, southern Brazil (Seixas 2000). Beyond the "living memory," oral history techniques can be used to access information on such events as major hurricanes, which occur about once per generation in parts of the Pacific (Lees and Bates 1990).

4.3.1.8 Short questionnaires

Short and issue-specific questionnaires can be useful for traditional knowledge research if conducted late in the research process. Once the researcher has an understanding of the local system, it may be feasible to use a questionnaire approach to quantify information such as the number of fish species identified by a group of fishers or the number of fishers using a lagoon or an area under consideration for an MPA.

4.3.2 Details of some selected techniques

The description of traditional ecological knowledge about fishery resources should include a folk taxonomy of fishery resources, a description of beliefs about important items in taxonomies, and a description of variation in ecological knowledge. Users' knowledge of the ecology can be obtained using ethnographic interview techniques (see Spradley 1969). The first step in acquisition of this type of information involves constructing folk taxonomies of fishery resources.

4.3.2.1 Folk taxonomies

Folk taxonomies for aquatic organisms such as fishes and marine invertebrates are best generated using a small group of experienced fishers. Since there is frequently a division of labour by age, gender, or some other criteria (for example, in some societies, females conduct inshore gleaning of invertebrates), this information must be obtained from representatives of the appropriate subgroups of the community. The first step is to ask them to name all the types of fishes they know that live on or around a particular habitat. The inquiry can be facilitated by asking informants to name organisms as observed at landing sites and markets. A picture book (colour pictures are best) can also be used to stimulate acquisition of fish names.

After this list is formed, the interviewer can then take each name on the list (for example, catfish) and ask if there are any other types of "catfish" locally. List construction will probably take several days, using about three hours of the fishers' leisure time each day. Ideally, the list should be cross-checked with another group, using the same techniques, but prompting with items from the first group if they are not in the final product of the second group. Similar methods can be used for other coastal and marine flora and fauna.

Scientific identification of taxonomic items can prove difficult. These lists are frequently surprisingly long. Pollnac (1980), using this technique in an examination of a coastal, small-scale fishery in Costa Rica, elicited 122 named categories of marine fishes captured by local fishers. For a coral reef in the Philippines, McManus *et al.* (1992) list over 500 species of fishes associated with a specific reef, and Pollnac and Gorospe (1998) list over 250 for a reef in another Philippine location. These findings suggest that reef fishers might have more complex taxonomies than the Costa Rican fishers in Pollnac's research. If someone with knowledge of reef fauna and flora taxonomy is present, he or she can attach the scientific nomenclature to the local name. If not, the researcher should take photographs (or collect samples) for later identification of species. Fish identification books, with colour photographs, can also be used as a supplementary method to link local and scientific names. Photographs are also an excellent stimulus for eliciting names. Where fishes change colour and characteristics with age and sex changes, the photographs should include representations of all stages. Some fishes also change colour when frightened and/or killed; these factors have to be taken into account.

In brief, the steps for conducting interviews to generate folk taxonomies are:
1. Identify user groups.
2. Using stimuli such as picture books and organisms in the wild (at landings and in the market), elicit names of fish.
3. For each type of fish named, ask if there are any other types of that fish.
4. Cross-validate information with additional informants.
5. Using fish (shellfish, etc.) identification books, identify fish by scientific name.
6. Photograph fish types that you cannot identify in the field so that experts in the university or fishery department may identify them.

4.3.2.2 Ethnographic information

For each (or each important) resource, investigators should elicit resource harvester knowledge concerning the resource. For example, for a given type of fish the investigator should question the harvester (or a group of harvesters, as discussed above for eliciting taxonomies) concerning numbers, locations, mobility patterns, feeding patterns, and reproduction. For each of these information categories, fishers should be queried concerning long-term changes. Reasons for changes should also be determined. Given the species diversity associated with coral reefs, this appears to be a formidable task, but such knowledge will probably be available only for important species. Those are the species the harvesters have been watching, hunting and eviscerating — the ones upon which most of their income depends.

Researchers should use ethnographic interviewing techniques to obtain this information. A good example of this type of information can be found in Johannes (1981), Lieber (1994), and Pollnac (1980, 1998). Questions that can be used to elicit this type of information for coastal fauna for each organism include:

1. Where is it usually caught?
2. Is it also caught in other areas?
3. Does the area change with time (hour, day, moon, month)?
4. In comparison with other organisms, what is the quantity available?
5. What other organisms are likely to be caught with it?
6. What does it eat?
7. Where and how does it breed?

It is important to note that there will probably be intracultural variation with respect to all aspects of traditional knowledge discussed above (Felt 1994; Berlin 1992; Pollnac 1974). Some of the variation will be related to division of labour in the community, as discussed above, but some will be related to degree of expertise, area of residence, fishing experience, and other factors. The conceptualization of "folk knowledge" as "shared knowledge" implies that care must be taken to not attribute idiosyncratic information as folk knowledge. This is difficult when using a rapid appraisal approach, especially given the anti-survey bias held by some ill-informed advocates of rapid rural appraisal. A survey of, say, 10 to 15 fishers concerning key aspects of "folk knowledge" can serve to rapidly identify areas of variability that could be addressed in planning future research for management purposes.

4.4.2.3 Cognitive mapping

Examples of cognitive mapping and variation in local knowledge are presented below (**Boxes 4.2** and **4.3**). Cognitive mapping is an aspect of local knowledge that is useful in determining knowledge about distribution of fish, breeding areas, and so on.

Box 4.2 Cognitive mapping.

One important aspect of local knowledge is user conceptualization of the distribution of the resource, including cognitive mapping. While distribution of a resource is a spatial phenomena, reference points in the spatial distribution are converted into concepts that are frequently named, especially if they are important reference points. Hence, users' cognitive maps of resource distribution can be constructed, in part, from named features, fishing spots, and so on. Place names elicited from the fishers of Discovery Bay, Jamaica, can be found in the figure below.

1. TOP CREEK
2. BOTTOM CREEK
3. SAMPSON POINT
4. PORT POINT
5. LONG BANK
6. TOP BAY
7. PUERTO SECO
8. DINA
9. OLD STATION
10. CLAUDY CAVE
11. MANTON BEACH
12. WHARF HEAD
13. JOHNSON ROCK
14. WHITE BEACH
15. SPAWN BAY
16. WHITEMAN
17. FRENCHMAN HOLE
18. LOBSTER POINT
19. FRY HOLE
20. MARINE LAB
21. LOWER REEF
22. LONGA SA
23. CHANNEL MOUTH
24. ONE BUSH

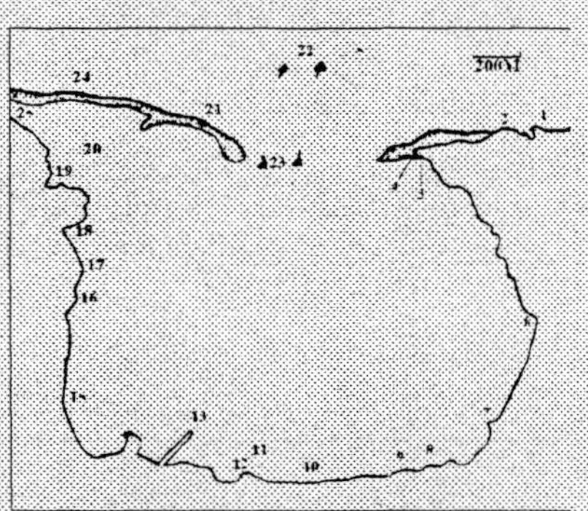

NOTE: NAMES TO THE EAST AND NAMES TO THE WEST ARE ORDERED TOWARD THE EAST OR TOWARD THE WEST AS ONE MOVES ALONG THE COAST. FOR EXAMPLE MOVING EAST FROM TOP CREEK, ONE FIRST ARRIVES AT AIRBASE POINT, THEN HOGFISH HOLE.

NAMES TO THE EAST
- AIRBASE POINT
- HOGFISH HOLE
- THROW OFF
- MACK MINUS
- DAIRY BLUE
- ONEHAND MAN
- GANJA POINT
- DAIRY POINT
- COCONUT WALK
- ROUND STONE

NAMES TO THE WEST
- SOLDIER WASH
- HEAVY SAND
- TURTLE SPOT
- LONGWALL POINT
- MISS RILEY BAY
- LOGON
- MUSCHET BEACH
- BLACK CAVE
- TRACER
- GRAVEL BAY
- LONDON BRIDGE
- RIO BUENO

The names of some places are obviously derived from shore-side structures or place names (for example, Marine Lab, Fort Point, Airbase Point), and others, significantly, reflect observation of fish behaviour at the place. For example, Soldier Wash is the place the fishers say the soldier crab comes to wash its eggs. Similarly, Spawn Bay is the place where fish spawn, and Fry Hole is a place where fish fry congregate. Some names refer to species found at the spot; for example, Hogfish Hole, Turtle Spot, and Lobster Point. Other place names are descriptive of actual features, such as Channel Mouth, Wharf Head, and Round Rock (adapted from Pollnac 1998).

Box 4.3 Resource-user traditional knowledge in the Philippines.

The fishers in Atulayan Bay, Lagonoy Gulf, the Philippines, have over 260 local names in their taxonomy of fish (Pollnac and Gorospe 1998). It is a truism in anthropology that cultural knowledge is unevenly distributed in any population; hence, one would expect intracultural variability in knowledge associated with a taxonomy as complex as the one used by the fishers of Atulayan Bay. Adequate investigation of this variability cannot be carried out within the time constraints of rapid appraisal, but an example illustrates the difficulties involved. The folk generic taxon *linhawan* is a good example. In an early stage of our research, an informant was queried concerning *maming*, a Labridae (wrasse). He called it a *linhawan*. He also classified other Labridae (for example, *talad*, *maming*, and *hipos*) as *linhawan*, but included *angol*, the hump-head parrotfish (*Bolbometopon muricatus*, a Scaridae) as a *linhawan*. A review of data collected several days previously, however, indicated that other informants identifying a picture of the hump-head parrotfish as *angol* sometimes use the Tagalog term *mulmol* for *linhawan*. In Tagalog, *mulmol* is identified as Scaridae. These informants noted that *linhawan*, other than *angol*, are classified by colour at the specific rank and gave the examples *linhawang asul* (blue), *puti'* (white), *dilaw* (yellow), and *itim* (black), all of which are Scaridae. Later informants added the folk-specific taxons *buskayan* and *tamumol* to the types of *linhawan* and denied that any of the Labridae are *linhawan*.

4.4 Literature acquisition and the Internet

4.4.1 Literature acquisition

Small and developing countries can find it quite difficult to access fisheries literature. In recent years, the number of fisheries-relevant publications has grown exponentially. Not only are there more journals dealing specifically with capture fisheries, but publications on law, policy, environment, coastal management, social sciences, and other disciplines more frequently contain fisheries articles as oceans, natural resources and their sustainable management become more topical issues. Still, none of the mainstream journals are specific to the management of small-scale fisheries. Academic journals are expensive, and fisheries authorities or universities in developing countries are seldom well endowed with them. If sought-after information is not in the refereed journals, chances are that it will be in technical or grey literature that is often even harder to obtain. Managers need to be selective while not letting the search process be too time-consuming.

However, a number of free or low-cost publications are readily available to most fisheries authorities, NGOs and other fisheries stakeholders. They vary considerably in content and are often associated with particular projects or programs. Some serve mainly to inform about and promote the activities or perspective of a particular organization. These biases should be noted, but once taken into account, the information contained is often useful to keep one up-to-date with international, regional, or local events and trends. **Table 4.1**, below, lists only a few of the many publications available. Readers should constantly be on the lookout for the emergence of new ones, as well as the termination of others due to project completion or funding constraints. Internet searches of the major international funding, donor, and project execution agencies often reveal additional sources.

TABLE 4.1 SOME NO- OR LOW-COST FISHERIES-RELATED PUBLICATIONS.

PUBLICATION*	PUBLISHER OR SOURCE
Food and Agriculture Organization (FAO) fisheries technical reports, papers, circulars, etc. report concisely on a wide range of fisheries matters.	Fisheries Department, Food and Agriculture Organization (FAO) of the United Nations (UN), Rome [www.fao.org/fi/]
OUT of the SHELL is a newsletter on coastal management, including fisheries.	Coastal Resources Research Network, International Development Research Centre (IDRC) of Canada [www.idrc.ca]
Samudra, a newsletter on fisheries and coastal zone issues, is also available online.	International Collective in Support of Fishworkers (ICSF), Madras [www.gmt2000.co.uk/apoints/icsf]
SEAFDEC Newsletter describes initiatives of SEAFDEC, a Southeast Asian treaty organization created in 1967 to promote fisheries development.	Southeast Asian Fisheries Development Center Secretariat, Bangkok, Thailand [www.seafdec.org.ph]
Seafish News informs about the activities of the Authority and fisheries developments in the UK.	Sea Fish Industry Authority, United Kingdom [www.seafish.co.uk]
EC Fisheries Cooperation Bulletin contains short technical and scientific articles about projects that are funded by the EU, often in developing countries.	Commission of the European Union, Brussels [europa.eu.int/comm/dg14]
SPC newsletters and information bulletins cover fisheries topics of interest to small island states.	Secretariat of the Pacific Community (SPC) [www.spc.organization.nc]
IIFET Newsletter promotes discussion of factors affecting international production of and trade in seafood, and fisheries policy questions.	International Institute of Fisheries Economics and Trade (IIFET), Secretariat, Oregon State University [www.orst.edu/Dept/IIFET]

TABLE 4.1 CONCLUDED

PUBLICATION*	PUBLISHER OR SOURCE
Sea Grant contains brief articles on marine projects and educational programs at several US universities.	For example: University of Puerto Rico Sea Grant College Program, Puerto Rico [www.nsgo.seagrant.org]
NAGA quarterly magazine features research and project summaries, news and notices of new publications, as well as upcoming workshops, conferences and symposia.	International Center for Living Aquatic Resources Management (ICLARM), Malaysia [www.cgiar.org/iclarm]
Common Property Resource Digest, which concerns all types of common property and efforts at research and management, includes publication lists and reviews.	International Association for the Study of Common Property (IASCP), USA [www.indiana.edu/~iascp/index.html]
SEACAM Newsletter addresses initiatives to assist the Eastern African coastal countries to implement and coordinate coastal management activities.	Secretariat for Eastern African Coastal Area Management, Mozambique. [www.seacam.mz/home.htm]
InterCoast is an international newsletter published three times each year to facilitate information exchange on coastal management.	Coastal Resources Management Project of the University of Rhode Island's Coastal Resources Center (CRC) [crc.uri.edu]
Ocean Update Newsletter is a monthly newsletter summarizing recent news, views and events concerning marine and coastal environments and wildlife.	SeaWeb is a project designed to raise awareness of the world ocean and the life within it. [www.seaweb.org]
AARM Newsletter informs about fisheries and aquaculture management in the Indo-China region.	AARM/Asian Institute of Technology, PO Box 4, Klong Luang, Pathumthani 12120, Thailand
Mekong Fish Catch and Culture informs on fisheries in the Mekong River region.	Mekong River Secretariat, PO Box 1112, Phnom Penh, Cambodia
Tambuli, a publication for coastal management practioners, includes information on coastal management	CRMP, 5th Floor, CIFC Towers, North Reclamation Area, Cebu City, Philippines [www.oneocean.org]
ContentsDirect, an e-mail service, provides the current contents of journals published by Elsevier.	Elsevier Science Ltd., Oxford, UK [www.elsevier.nl/locate/ContentsDirect]

* Listing here is for example only, and does not imply any endorsement of the publication or publisher. Descriptions are taken from the organization's Web site, where available. Each address is preceded by " http:// "

One simple way to keep track of the contents of many journals is to subscribe to their free print or e-mail services that disseminate content tables and abstracts. Users of these services may use e-mail to request reprints directly from authors, or may obtain copies of articles when traveling to destinations with major libraries. University and other major libraries tend now to be accessible via remote computer connections that allow the user to browse their collections with links to several national and international databases. Databases such as those of the Food and Agriculture Organization (FAO), Department of Fisheries and Oceans (DFO) of Canada, European Commission, National Marine Fisheries Service (NMFS) of the USA and others can be also queried directly to reveal a surprising amount of information on a wide variety of topics.

4.4.2 The Internet

Because the sources above largely depend on having access to computers and reliable Internet providers, we advise fishery workers, researchers, and managers to invest in obtaining access to Internet resources. Used judiciously, with practice an enormous amount of useful information is at your fingertips. In many ways it is access, or lack of access, to this resource that will make a significant difference in the cost of data or information acquisition and thus reduce or increase the gap between the developed and developing countries. The Internet also provides a cost-effective means of communicating to many sources of information, even if the information itself is not easily or freely accessible in a public domain.

The Internet offers the small-scale fisheries manager a tremendous opportunity to quickly obtain access to vast amounts of fisheries-related information — some of it useful, some not. However, with Internet access, the connected computer is an invaluable tool for communication, information exchange, and research that no small-scale fisheries manager should overlook. Compared with using telephone, fax or airmail, electronic mail (e-mail) is usually more cost-effective. For information on e-mail services, contact a recommended Internet provider. Look for the electronic addresses of organizations in their letterheads or documents and those of individuals in lists of meeting participants. Internet directories in general, or organization-specific ones, provide lists of addresses similar to the way a telephone book does.

The World Wide Web provides access to a variety of resources that would be otherwise unavailable in some locations or expensive in time or money to obtain. Fisheries- and coastal management-related sites on the Internet are named or reviewed in some of the low- or no-cost publications listed in **Table 4.1.** See, for example, *Out of the Shell*, the Coastal Resources Research Network Newsletter.

Readers not familiar with the Internet may find it useful to consult an introductory guide on the subject. Online tutorials are now available with most Internet software, so the beginner will probably not have to invest in "how-to" manuals. Since Web sites and e-mail addresses change, the reader may occasionally need to update the following sources. Those listed in **Table 4.2** are primarily locations with several links to other important information resources.

TABLE 4.2 ANNOTATED LIST OF SOME FISHERIES-RELATED INTERNET WEB SITES.

ORGANIZATION DESCRIPTION AND INTERNET ADDRESS*
American Fisheries Society (AFS), which promotes scientific research, management and education for fisheries scientists, publishes fisheries research journals. [www.fisheries.org]
World Wide Fund For Nature is the world's largest and most experienced independent conservation organization. [www.panda.org]
Aquatic Network concerns aquaculture, fisheries, ocean engineering, marine science, seafood, etc. [www.aquanet.com]
Center for Marine Conservation (CMC), USA works to conserve the abundance and diversity of marine life and protect the health of oceans. [www.cmc-ocean.org]
Centre for the Economics and Management of Aquatic Resources (CENMARE), UK provides fisheries economics links, research papers, and academic training. [www.pbs.port.ac.uk/econ/cenmare]
Department of Fisheries and Oceans (DFO), Canada provides information on Canadian federal fisheries issues and activities. [www.pac.dfo-mpo.gc.ca] and [www.mar.dfo-mpo.gc.ca/e]
European Commission DGXIV – Fisheries addresses Common Fisheries Policy, the EU work program and overseas projects. [europa.eu.int/comm/dg14]
FishBase global information system on fishes is a relational database with fish information catering to scientific professionals. [www.fishbase.org]
Fish-Link provides information on aquaculture and fisheries. [www.fishlink.com]
Food and Agriculture Organisation (FAO) Fisheries Department, Rome addresses FAO programs, projects, documents, databases, data atlases, and much more. [www.fao.org/fi]
Gadus Associates, Nova Scotia, Canada maintains numerous links to sources of information on marine commercial fisheries. [home.istar.ca/~gadus]
IFREMER (the French institute for research and exploitation of the sea) describes French marine and fisheries-related projects, consultants, research capability, and vessels. [www.ifremer.fr/anglais/institut/index.htm]
Intergovernmental Oceanographic Commission (IOC) of UNESCO deals with oceanographic research and marine science generally. [ioc.unesco.org/iocweb]
International Collective in Support of Fishworkers (ICSF), Madras is a global NGO network of resource persons, projects, materials, and activities. [www.gmt2000.co.uk/apoints/icsf]
International Commission for the Conservation of Atlantic Tunas (ICCAT) is responsible for international management of tunas and tuna-like species, species assessments, and related meetings. [www.iccat.es]
International Council for the Exploration of the Sea (ICES) is a scientific forum for the exchange of information on the sea, its living resources, and marine research. [www.ices.dk/toc.htm]
International Development Research Centre (IDRC), Canada provides information on its organization and programs. [www.idrc.ca]
International Institute of Fisheries Economics and Trade (IIFET) addresses aquaculture, marine resource economics, trade in seafood, use of marine resources, and fisheries management. [www.orst.edu/Dept/IIFET]

(continued)

Table 4.2 Concluded

Organization description and Internet address*
World Conservation Union (IUCN) is the world's largest conservation-related organization. Its mission is to influence, encourage and assist societies throughout the world to conserve the integrity and diversity of nature and to ensure that any use of natural resources is equitable and ecologically sustainable. [www.iucn.org]
National Marine Fisheries Service (NMFS), USA is the source for US fisheries databases, new policies and laws, and fisheries science papers. [www.nmfs.gov]
Natural Resources Management Unit (NRMU) of the Organization of Eastern Caribbean States (OECS) deals with fisheries, the coastal zone and environmental matters in the eastern Caribbean islands. [www.oecsnrmu.org]
Secretariat of the Pacific Community (SPC), New Caledonia has programs, publications, directories and Pacific area fisheries information. [www.spc.organization.nc]
United Nations Division for Ocean Affairs and the Law of the Sea (DOALOS) is a central source for information on international marine law. [www.un.org/Depts/los/index.htm]
World Forum of Fish Workers and Fish Harvesters, India is a collective action NGO that relates to rights-based fisheries information and advocacy [www.south-asian-initiative.organization/wff/intro.htm]
US Fishery Management Councils are eight regional fishery councils that manage the living marine resources within the US exclusive economic zone, which ranges between 3 and 200 miles offshore. [www.noaa.gov/nmfs/councils.html]

* Listing here is for example only, and does not imply any endorsement of the publication or publisher. Descriptions are taken from the organization's Web site, where available. Each address is preceded by " http:// "

Other Internet resources include electronic mailing lists wherein the user subscribes, often for free, and is then part of a virtual community exchanging information on topics of common interest. Universities or government agencies often maintain the lists, and some have a moderator who screens content. On the unmoderated sites anything can be posted, subject only to good etiquette (called "netiquette" on the Internet) and written sanctions by fellow users. It is quite acceptable to be a "lurker" on mailing lists, following the discussions without feeling compelled to contribute. Mailing lists such as "FISHFOLK" are an excellent way to keep abreast of current issues and make contact with others sharing this interest, including experts in the field. Questions or requests for information posted onto a list often yield rewarding results.

Newsgroups are another means of electronic information exchange, performing a function similar to a bulletin board for posting information, or making requests for information. Of the thousands of newsgroups on nearly as many topics, a few are fisheries-related. Internet software allows you to browse the entire list of newsgroups and select the ones that you wish to subscribe to. Your computer then receives the postings to that group when you request them. Often, dialogues or group discussions are taking place on some topic, similar to the discussions on an electronic mailing list.

It is worth learning how to perform Internet searches. By typing in key search terms, such as titles or names, the Internet user can discover an abundance of Web sites, lists, and e-mail addresses related to the topic of interest. Search engines are also very useful for finding sites that have changed address or items on which you have no information other than a name or acronym.

4.5 Analysis and Interpretation

Data and information usually require some analysis and interpretation before they are useful for a particular application. The increasing availability of easy-to-use computer software is an asset to fisheries managers. Several sophisticated analyses can now be done on any moderately powerful personal computer, often using basic spreadsheet and statistical applications. With some training and practice, it is not difficult to formulate and use your own analytical routines. Fisheries-specific software also exists, mainly for stock assessment (for example, FISAT). Nonetheless, basic exploratory data analysis, especially using graphical methods, may prove more versatile and valuable than specialized fishery software. Uses include identification of simple trends, cycles and other patterns in the data.

More attention is now being paid to attaining a basic understanding of trends and changes affecting fisheries and less on highly quantitative models. Reasons for this include the issue of uncertainty dealt with in Chapter 2, the need to include human dimensions that are less easily quantified within the same models, and the generally disappointing experiences worldwide of reliance on quantitative approaches (Wilson *et al.* 1994; Ludwig *et al.* 1993).

Desktop analysis based on information from other, similar areas is often a valuable way to make a preliminary assessment of the major problems in a fishery. The Jamaica Pedro Bank conch fishery is an example of a fishery where quotas were set using empirical yield estimates from other areas (Aiken *et al.* 1999).

For small-scale fisheries management, a matter to pay attention to is the use of group processes in information analysis as a means of promoting stakeholder participation. Focus groups and methods such as the participatory processes described in the previous chapter are relevant. The use of graphic, highly visual techniques for analysis, particularly as inputs to stakeholder interpretation, can be advantageous for several reasons. These include leveling the playing field so that those used to conducting sophisticated analyses are not necessarily assigned superiority over people who deal daily with the fishery resource and fishery workers. If thoughtfully constructed, graphics can simplify the tasks of analysis and interpretation without losing the power of explanation and prediction sought in models.

Some may argue that this task of focusing on the main issues or components can itself be enlightening, especially in the multidisciplinary approaches advocated for involving diverse stakeholders. Examination of the system used in the USA, for example, reveals that resource users and industry representatives often hold widely divergent interpretations of data, particularly if the data are presented as indisputable fact (McCay and Wilson 1998). These non-scientists see the processes of analysis and interpretation as deeply flawed, suspecting biases toward the results favoured by other, more powerful, stakeholders. This is almost inevitable in the absence of participation in these processes, but it may not be completely dispelled by inclusion. Respect and trust among all stakeholders play a large part in the success of participation in this stage of the information system, especially where specialized skills are required, such as statistical analysis or distinguishing between cause and effect or coincidence in correlated variables.

4.6 Information Management

Once information is generated, attention should be paid to its management, particularly through the use of electronic and automated technologies, where feasible. Fisheries managers, who are also administrators, should be familiar with the usual means of organizational record-keeping in paper files, ledgers, subsidy or loan applications, accounts books, and the like stored in safes, shelves, and filing cabinets. Regulation may require that several of these be kept and archived for a particular number of years. Some of these ordinary records will likely be relevant to fisheries management. The manager should identify those records and make sure they remain accessible, especially if a time-series of information, such as changes in employment, engine sizes or boat lengths, will be needed for future analysis. Crucial paper records may be microfilmed or put on microfiche. (See also Chapter 3 on documentation).

Legal and administrative documents are usually stored as paper records, but daily operations are often performed with their electronic versions. These versions, plus the biological, social, economic, and fishery assessment information, should be backed up in case of computer problems. There are several media available for backup storage, and this can usually be automated. Storage on compact disc, one of the most durable options, is becoming increasingly affordable. The weak link is the human one: ensuring that the backup system is well planned and executed. While this is stating the obvious, we know of many cases in which critical data and information have been lost due to the absence of these simple systems. This also links to the matters of institutional memory and historical analyses (**Box 4.4**).

Small, in-house libraries are the norm in many fisheries authorities, and managers should be aware of readily available bibliographic software that facilitates managing libraries, large or small. Reference management applications can be used to insert citations directly into articles being written. Given the introductory chapter's exhortation for small-scale fishery managers to put more in print and distribute to both national and international readers, these applications should be of interest. Most word processing software makes the publication of professional-looking documents easy. It may be important to keep a special collection of these nationally generated documents, as well as any other exceptionally useful ones, since these could be in demand for a long time. Managers may wish to establish different levels of access to library materials and borrowing or reading privileges. Photocopying of material, a

BOX 4.4 INSTITUTIONAL MEMORY.

Many studies of traditional ecological knowledge show the key role that elders play. They act as keepers of ecological knowledge, they help transmit knowledge by direct teaching and through rituals and oral history, and they provide the wisdom to interpret novel observations. In short, elders in traditional societies are the main agents of institutional learning. By contrast, modern society and resource management agencies do not rely on elders; emphasis is put on new knowledge and youth. Nevertheless, some are beginning to think that elders may have an important role to play in resource management.

The mechanism for institutional learning, as for any learning, is trial-and-error. International experience with resource management agencies shows that there often is potential for institutional learning following a crisis. But many large institutions seem unable to convert experience into learning that leads to more effective resource management. Hilborn has suggested that this may be due to poor institutional memory. The steps in learning from experience include documenting decisions, evaluating results, and responding to evaluation. But also needed are mechanisms of institutional memory to retain the lessons learned. Publications, data records, and computer databases are often not adequate to serve as the institutional memory. As Hilborn puts it, "The richest form of memory is stored in the cerebrum of the staff of fisheries agencies. We sometimes forget how much an individual may have learned in 20, 30, or even 40 years of work in an agency. For each documented experience, there are probably 10 that are left unwritten. Those that are documented may be a biased sample. Journals do not often publish negative results; managers don't like to hear bad news — we don't document our failures. When someone retires, much information walks out of the door along with the gold watch."

Source: R. Hilborn 1992. "Can fisheries agencies learn from experience?" Fisheries 17 (4): 6–14.

matter for the applicable copyright policy or law, is an issue of intellectual property that should be treated seriously. The payment for materials should also be governed by policy, perhaps making distinctions between local resource users or students versus overseas academic researchers or consultants. Some agencies' Web sites list and describe publications that they distribute free or for a price. Controlling access to information is a necessary dimension of its management.

Geographic Information Systems (GIS), becoming more popular as management tools in several different forms, have potential for numerous fisheries applications (Meaden and Do Chi 1996). They facilitate spatial management and display geographically referenced information in uses that range from habitat mapping to catch and effort logging to the compliance tools of Vessel Monitoring Systems (VMS). The latter allow shore stations to track the movements of fishing vessels fitted with special transponders via satellite. This automatic relay of information facilitates monitoring, control, and surveillance. Several of these systems are fairly simple and affordable once the requisite computer hardware is available. Since most people, particularly resource users and policymakers are familiar with the geography of their country or region, the innovative management of geographic systems is useful for data acquisition, analysis, interpretation, and communication. Mapping techniques, some of which have been mentioned previously in this chapter, are an important means of storing and communicating fishery information. Butler *et al.* (1986) provide an overview of the use of mapping techniques for fisheries.

4.7 Communication and Use of Information

While managing access to information is important, even more critical is its active communication to stakeholders. As mentioned in the previous chapter on the planning process, not all stakeholders will seek out information unless it is marketed in a way that meets their needs, perhaps after assisting them to identify those needs. Managers must communicate information in a form that is appropriate for each audience, making information available to stakeholders so they can participate in an informed way (**Figure 4.1**).

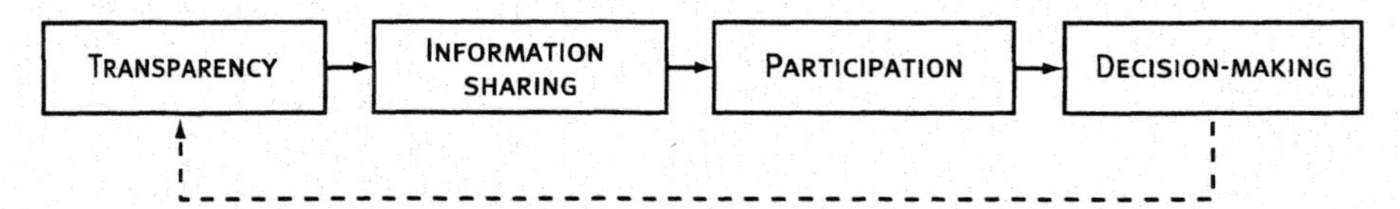

Figure 4.1 Self-reinforcing information system.

In today's world of high technology and powerful marketing tools, the small-scale fisheries manager needs to be aware of the importance of properly conceived and executed presentations: written, oral, visual, or a combination of these. High-level decision-makers may have become accustomed to professional presentations from consultants and external agencies and at international conferences.

It is relatively easy for the small-scale fishery manager to make a sophisticated presentation. Even though appearances can be deceiving, it is prudent for the fishery professional to make every effort to communicate his or her ideas in the most effective manner appropriate to the situation and available presentation technology (**Box 4.5**).

The manager must also look at the levels of decision that the information may be

Box 4.5 Some pointers on presenting information.

Though not intended as a presentation manual, this box gives pointers on how to make effective presentations, concluding with sources of further information. Many resources about making presentations are available via the Internet. The following lists draw upon several of those sources, as well as upon the authors' experiences.

Reports and large documents

- Put a cover on your document — introduce its purpose and give it a polished look.
- Provide citation information — date, author, originating organization and address.
- Include a table of contents — help readers navigate its contents.
- Add an executive summary — provide the key points to those who do not read the entire document.
- Use short paragraphs — give readers continuous breaks along the way.
- Choose a 12-point typeface — for optimum readability and reproduction quality.
- Insert a line-space between each paragraph — improve readability and reduce fatigue.
- Put key information as bulleted text, charts or diagrams — help readers quickly absorb it.
- Avoid use of jargon when possible, and explain any jargon that you can't avoid — help readers to understand.
- Include graphics, charts and diagrams — relieve text-heavy pages.
- Consider whether footnotes or endnotes are useful — they have their pros and cons.
- Enhance headlines, subheads and other small pieces of text — relieve the eye.
- Keep font and other hierarchies consistent throughout — improve document appearance.
- Use "styles" in word processing — add consistency and make routine changes easier.
- Tailor document length and layout to the audience — think of marketing the information.
- Consult texts such as style manuals, dictionaries, etc. — they improve your writing.
- Use proper citation for author, date, title, and publisher — help reader to find references.
- Search the Internet for online writing resources — they are useful if not overused.
- Check samples of articles if writing for a journal — this supplements instructions to writers.
- Remember to update backup copies of your work — technology fails when you least expect it.

Computer, overhead and slide presentations

- Remember to use the KISS principle — Keep It Short and Sweet.
- Present text in summary points — preferably bulleted points, not entire sentences.
- Limit the amount of information on a slide — five messages or fewer is better.
- Print all text in 18-point size or larger, depending on the projection arrangements.
- Limit the number of font typefaces to two clearly readable ones.
- Use bold, italic, or colour to provide visual emphasis instead of underlining text.
- DO NOT USE ALL UPPER CASE IN LONG STRINGS OF TEXT. IT IS HARD TO READ, GIVES THE IMPRESSION THAT YOU ARE ALWAYS SHOUTING AT YOUR AUDIENCE, AND LEAVES NO ROOM FOR EMPHASIS.

- Choose dark backgrounds and light-coloured text; this generally works better than other combinations.
- Use cool colours for backgrounds — they tend to recede from the eye.
- Use high contrast for readability and legibility.
- Be consistent in formats, fonts, and other repetitive features throughout the presentation.
- Bigger is better, especially for presentations in large rooms.
- Use only the most appropriate charts and graphs — those that require the least explanation.
- Use Clip Art graphics and illustrations for a good reason, not just to decorate. Clip Art can communicate a point, capture attention or lighten up a heavy subject.
- To keep Clip Art effective, don't put it on every slide.
- Take a variety of media options to the presentation — if the computer projector breaks down, move immediately to the overhead slide version or printout of the presentation that you have on standby.
- Test the presentation — ideally, at the location where you will be presenting it.
- If you are new to computer presentations, practise how to return to a slide, move ahead to a particular slide, and to stop the presentation.
- For presentation applications, establish three to five different templates that work well for you, and be prepared to switch templates if lighting, projection or other problems arise with a specific template.
- Keep a sense of humour; presenters that look pained will pain the audience.

FURTHER READING AND INTERNET SOURCES

(a) In print

Bandy, G. 1993. *Multimedia Presentation Design for the Uninitiated*. Aldus Corporation.

Parker, R. C. 1993. *Looking Good in Print*, 3rd edition. Ventana Press.

Robinette, M. 1995. *Mac Multimedia for Teachers*. Foster City, CA: IDG Books Worldwide.

Siebert, L. and L. Ballard. 1992. *Making a Good Layout*. North Light Books.

Siebert, L. and M. Cropper. 1993. *Working with Words & Pictures*. North Light Books.

(b) Online

The Writing Center [www.researchpaper.com/writing.html]

The Writing Process (University of Texas) [uwc.fac.utexas.edu/fac/aboutswc/man/process.html]

Webster Dictionary Search Page and Thesaurus Search Page [www.m-w.com/dictionary.htm]

Writing Style (University of Virginia) [www.lib.virginia.edu/reference/index.html]

used for, and tailor the information to the decision-making environment. For example, providing decision-makers with convincing information may often require the compilation of information at a level that is too general for use in management. This type of compilation is an important part of fishery assessment. The preliminary assessment of the gross contribution of fisheries to the economy of Antigua and Barbuda is an example of the level of compilation that is needed at the national planning level to support a request for funds needed for management (**Box 4.6**). This assessment also underscores the importance of developing policies to ensure that the tourism industry is adequately supplied with fishery product, since this is where most of the added value comes from. This may increase the number of groups interested in fisheries information.

Box 4.6 Valuation of fisheries in Antigua and Barbuda.

Antigua and Barbuda is a small, two-island state in the northeastern Caribbean. About 650 fishers operate about 355 vessels, harvesting demersal resources (reef fishes, conch, and lobster) on the island shelf. Seasonal pelagic fish harvest is almost entirely recreational. Fish vendors sell to consumers and restaurants, most of which serve tourism. There are several exporters of fish, lobster, and conch. There was no estimate of the value of the fisheries that incorporated value added by processing, export and restaurant retail.

A preliminary analysis of total gross revenue from the fishery sector took about two days, using existing information. A path diagram for the flows of fishery products from producer to consumer was constructed from the knowledge of fishery officers. Information on prices came from the Fisheries Division. Hitherto, the value placed on fisheries was landed value only. The new estimate indicates that the value-added component of the gross value of the fishery is substantial, being 65.1 percent in 1994 and 66.4 percent in 1995. This is largely due to the use of fishery products in tourism, and is highest for lobster and conch, which are high-priced restaurant items. There are no data from which to estimate the contribution of recreational fishing, particularly from charter boats, but a preliminary estimate was derived. There are about 20 cruisers available for charter trips, which cost about US$400. If each vessel operates for 40 weeks and makes three trips per week, the gross revenue from this subsector would be US$960 000 per year.

More detailed estimates of the value of fisheries in Antigua and Barbuda would be useful. However, the preliminary estimates provided here enable decision-makers to determine how important fisheries are. It indicates that the contribution of fisheries to the national GDP in 1995 was about 6.5 percent. This is in contrast to the official value of 4 percent for the agriculture and fisheries sectors combined, of which about 45 percent (1.8 percent of GDP) is considered to be due to fisheries.

GROSS VALUE OF FISHERIES PRODUCTS, INCLUDING VALUE ADDED BY EXPORT, LOCAL SALES BY MIDDLEMEN, AND RESTAURANT SALES.

Item	Lobster		Conch		Finfisk		Total	
	1994	1995	1994	1995	1994	1995	1994	1995
Total Landing (mt)	222	317	208	153	1 462	1 487	1 892	1 958
Ex vessel price/kg	8.15	8.15	3.46	3.46	4.07	4.07		
Total ex vessel value	1 807 356	2 583 967	719 248	530 735	5 957 917	6 059 572	8 484 520	9 174 274
Percent exported	2.10	2.56	0.24	0.00	7.50	9.80		
Total exports (mt)	4.7	8.1	0.5	0.0	109.7	145.8		
Export price/kg	24.20	24.20	5.50	6.60	5.65	5.65		
Value added by exports	74 770	130 315	1 015	0	172 847	229 707	248 633	360 022
Amount sold to rest. &. cons. (mt)	211	291	207	153	1 353	1 342		
Price/kg to rest. & cons.2	11.77	11.77	4.06	4.06	6.49	6.49		
Value added by local sales	766 206	1 055 683	123 187	91 355	3 274 106	3 247 170	4 163 498	4 394 208
Percent sold to restaurants	90.00	90.00	30.00	30.00	40.00	40.00		
Amount sold to restaurants (mt)	200	285	62	46	585	595		
Price/kg as meal in restaurant	39.72	39.72	33.33	33.33	13.33	13.33		
Value added by restaurant sales	5 579 877	7 977 520	1 824 057	1 345 977	4 000 458	4 068 715	11 404 392	13 392 213
Total	8 228 208	11 747 484	2 667 507	1 968 068	13 405 328	13 605 165	24 301 044	27 320 717
Value added as % of total value	78.0	78.0	73.0	73.0	55.6	55.5	65.1	66.4

4.8 Conclusions

This chapter provides just an overview of some of the considerations to be taken into account when establishing or operating fisheries information systems. The extremely large literature that exists on information technology and management should be consulted for detail. The most important point to remember is that both information and lack of information have costs. Managers must weigh their information needs against other expenditures. Moreover, they must carefully design their information systems in order to get the most out of them. The greater the number of potential end users and decision-making situations, the greater is the attention that must be paid to these choices. The next chapter continues to address information, this time in the context of evaluation.

Chapter 5
Project Assessment and Evaluation

5.1 Introduction

Previous chapters presented an alternative way of looking at the problem of fisheries management. Arguments are made for an approach that is more complex than the usual resource assessment. It is complicated by arguments that the entire fishery, including the fishers and their communities (that is, the stakeholders) must be assessed. Such an approach requires a different type of information to set, meet, and evaluate the achievement of the new management objectives. This chapter outlines a framework and process for obtaining this type of information.

We are looking at the fishery management process as one that involves a preliminary appraisal that is used to set objectives as well as to develop the strategies and tactics for achieving the objectives. From this perspective, we can view the process as a project. Projects must be monitored if they are to be kept on track, and evaluated if we are to learn from our successes and failures. Adequate monitoring of fishery management projects also allows us to fine-tune project strategies and tactics to more effectively respond to both environmental and stakeholder impacts. In many cases it is practical to start small — to test the selected management strategies and tactics in a relatively small, easily monitored area. In other words, it is prudent to start with a pilot project — to learn from successes and failures in a relatively small area before expanding to a larger area, especially with a novel approach. It is easier to correct errors in a small area than a large area, and a success in a small area can be used as a demonstration site for a larger scale project. The area selected for the project is referred to as the "target area."

Box 5.1 The distinction between project objectives, strategies and tactics.

Let us say that one of the objectives identified is to increase coral reef fish populations. A strategy for achieving this objective may be to establish marine protected areas over sections of heavily exploited and degraded coral reefs. Tactics for implementing the strategy might include: 1) hold fisher community meetings (and use other educational techniques such as posters) to explain the concept of marine protected areas, 2) identify stakeholders, 3) involve fishers in assessment of coral reefs to select appropriate areas, 4) involve stakeholders to develop appropriate regulations, 5) develop and provide appropriate surveillance and enforcement mechanisms, and 6) deploy marker buoys around the perimeter of the protected area.

Box 5.2 Project area is directly related to the scale of management.

Pilot projects frequently target a small area, such as a village, several villages along a designated stretch of coastline, a bay, or a gulf. Sometimes a management project (not a "pilot project") can target such small areas. Fishery management projects can also target larger regions (for example, combinations of states or provinces) or the entire nation. The described logic of project assessment, monitoring, and evaluation applies equally well to all scales of management: the quantity of information increases as the scale increases.

In this chapter we first outline the general phases and logic associated with the project cycle — the process of assessing, monitoring, and evaluating a fishery management project. Then we briefly describe the steps to take. Information is necessary at all stages of the project for setting of objectives, developing strategies, and continually monitoring impacts to adapt the process as well as to learn from failures and successes. Then, a section of this chapter outlines the information needed for management of a fishery with community involvement, based on variables discussed in previous chapters and a review of the literature. The Appendix provides detailed information about methods of data collection.

Box 5.3 One can define a "community" in several ways.

Jentoft et al. (1998) make a contrast between a functional community and a local community. The former can refer to groups defined by fishing grounds, gear types, or species. The latter, the local community, is defined principally in terms of place or location where a group of interacting people are tied together by residence, identity, and history (for example, a village or town). Much of our emphasis here is on the local community as being the primary unit of management. This does not exclude, however, the functional community, which might, in some circumstances, be the focus of the management effort (see Chapter 3). In this case, one must realize that members of the functional community are also members of local communities, which are the context of their every day lives and thus affect the attitudes, beliefs, and values that influence their fishery-related behaviour.

5.2 The information stairway

The process and methods described in this chapter are designed to be applied at four distinct phases of a fishery management project. We can think of these phases as steps on the information stairway to effective fishery management. The steps are: 1) preliminary appraisal, 2) baseline assessment, 3) monitoring, and 4) evaluation.

Information for each of these steps is essential to maximize chances that the project will be effective for the adaptive management process and to acquire the learning needed to apply the process in other regions.

5.2.1 Preliminary appraisal — what do we do first?

The preliminary appraisal provides information from the target area to be used in preliminary identification of management objectives as well as pilot-project site selection. The target area is selected on the basis of its being a definable area (such as a section of coastline or a politically defined region) with identified or potential resource problems. As discussed in chapters 3 and 8, there are obvious preconditions to successful local-level or comanagement of fishery resources, and the objective driven management process requires basic information on the fishery to identify objectives and the strategies and tactics appropriate to achievement of the objectives. Hence, preliminary information is essential for identifying fishery resource problems as well as locations with characteristics that suggest they would be amenable to management strategies appropriate to the manager's resources. Careful attention to the identification of such locations is necessary to increase the probability of project success. These successful sites can then be used as demonstration sites for visitors from other areas, as well as for justification for further funding.

5.2.2 Baseline — where are we starting from?

The baseline provides detailed information on the pilot project sites, as well as information essential for monitoring and evaluating project impacts. The information obtained includes indicators essential to objective driven management of the fishery (Chapter 6), as well as those related to human aspects of local and/or comanagement of the resource. A number of variables have been identified as associated with achievement of project objectives (see previous chapters and World Bank 1999; Pomeroy *et al.* 1999, 1997; Novaczek and Harkes 1998; Pollnac and Pomeroy 1996); these should be assessed as part of the baseline. These variables form part of the context of the project. Objectives of the management process (for example, fishery ecosystem health, including the humans in the fishing communities) are conditions for which impact variables must be developed and assessed during the baseline. Concurrent with the collection of information on these indicators at the pilot project sites, it is necessary to collect a subset of this information (indicators of project objectives such as "health" of the fishery) from nearby areas being used as control sites.

Box 5.4 Project variables.

These are the specific tactics and strategies (see **Box 5.1**) identified to achieve project objectives. All the other factors that can influence both implementation and impact of the project strategies and tactics are classified as context variables. These, which include natural environmental factors that may result in the failure of a given tactic, social variables that could result in conflict over perceived impacts, and so on, are detailed in the section on information needs.

Box 5.5 Control sites.

These are areas (communities, stretches of coastline, etc.) where project strategies and tactics will not be implemented. They are used to determine if the project itself caused any observed effects or if some other, non-project, variables influenced the results. While it would be unrealistic to assume that a true experimental design (for example, random assignment of project and control sites) could be applied, a quasi-experimental design (nonrandom assignment of project and control sites) is feasible. This option is preferable to a simple before and after comparison of the project sites (the interrupted time-series design, see Pomeroy 1989), which would not be as effective for evaluating alternative (that is, nonproject variable) explanations for observed impacts.

Control sites are important for several reasons. A significant question associated with any fishery management project is its effect on the human and non-human components of the fishery ecosystem. Ideally, both of these components will benefit from a fishery management project. The only way we can determine these effects, however, is by establishing a baseline composed of indicators that can be compared with similar data collected during and after the project. Obviously, however, a fishery can be affected by factors other than those generated by a management strategy. Outside forces, both natural and unnatural, can affect a fishery ecosystem: think of changes in weather patterns, in infrastructure, and in the social, political, and economic context of the involved communities. Therefore, in addition to baseline information, it is necessary to collect information from similar areas to use as a control to determine what has influenced the ecosystem: the introduced fishery management strategy or some other factors.

5.2.3 Monitoring — what is happening?

Monitoring is one of the most important aspects of project implementation. Through monitoring, project managers learn if fishery management interventions are working. If not, they learn what is influencing achievement of objectives: the project activities or some contextual variable. With this learning, they can adapt project activities to better fit the existing situation. This approach, called adaptive management (cf. Margoluis and Salafsky 1998), is the way to learn from mistakes so that one is not be doomed to repeat them — as so often happens in fishery management projects.

Monitoring keeps track of implementation activities as well as their effects on achievement of the ultimate objective — a healthy fishery. One can conceptualize each of the project activities as various levels of objectives beneath the ultimate objective or purpose. For example, an intermediate objective may be to establish an MPA (see the project strategy in **Box 5.1**). This would be accomplished by establishing and implementing sub-objectives, such as community consultation, identification of boundaries for an MPA, and development of institutions for surveillance and enforcement (see the project tactics in **Box 5.1**). The monitoring process is designed to keep track of the degree to which these various levels of objectives are achieved. The baselines for both project and control communities provide the standard for comparison and information on the pre-project status of the fishery and related variables. During monitoring, the baseline information is compared with that collected during project implementation.

The logic of the monitoring and evaluation system proposed here is as follows. A project management team develops a set of intermediate objectives aimed at improving the ecosystem's well-being, then implements activities to achieve the intermediate and ultimate objectives. Each step in this process involves decisions and activities that can influence achievement of objectives. A wide range of variables are associated with achievement of project objectives (see previous chapters, World Bank 1999; Pomeroy *et al.* 1999, 1997; Novaczek and Harkes 1998; Pollnac and Pomeroy 1996). To learn from the process of monitoring and evaluation, we must account for those variables that may be responsible for observed levels of achievement of objectives.

Conceptually, variables that are expected to influence the level of achievement of fishery management objectives are classified as independent variables. These variables are further subclassified as project variables and context variables (see **Box 5.4**). Project variables include aspects of fishery management planning and implementation; context variables are the non-project, independent variables. Examples of context variables are the social, political, and economic aspects of the larger context of the project ecosystem — the community or communities and their natural environments that are included in the management project — (for example, resource management–related national legislation, markets for project ecosystem products), as well as that ecosystem's techno-economic, biophysical, and sociocultural aspects (for example, technology used in harvesting the resource, value of resource for income and household nutrition, perceptions of resource abundance, natural boundaries of resource, local ecological knowledge, and level of community cohesion).

The dependent variables (impact variables) that indicate the degree of achievement of the various levels of fishery management objectives constitute the second category of variables. These are composed of two subsets: 1) achievement of intermediate objectives and 2) effects on the fishery's well-being. The first includes consideration of the degree of achievement of both material (for example, trap mesh size changes, construction of meeting and information centres) and non-material (for example, training, institution building) objectives. The second subset considers project influences on the fishery, including separate measures of the human and natural components. Each of these components is, in turn, composed of distinct sets of variables (for example, changes in income, access to resources, availability of resources). Although categorized as dependent variables for one level of analysis, the achievement of intermediate objectives can also be conceptualized as antecedent to achievement of the ultimate objective of fishery ecosystem well-being.

Ideally, the monitoring process begins as soon as project activities directed at achievement of intermediate and ultimate objectives begin. It may not be realistic to expect measurable progress toward all ultimate objectives, especially those related to quantity of fish, in a project's first few years. One can, however, assess project activities and changes in contextual variables and possibly identify issues missed in baseline assessments, as well as evaluate achievement of intermediate objectives.

5.2.4 Evaluation — what actually happened?

Two types of evaluation are necessary. Post evaluation is conducted soon after all project activities have been carried out; ex-post evaluation is done several years following those activities (see Morrissey 1989). The logic of the process is similar to that of monitoring — levels of achievement of intermediate and ultimate objectives are evaluated along with non-project variables (for example, the independent and dependent variable sets described above). At this point, the data are compared with those collected as part of the baseline to determine degree of management project impact. At both of the evaluation stages, the control sites must be evaluated with respect to indicators reflecting the health of the fishery ecosystem as well as other fishery-related variables that could change the health of the fishery (for example, the context variables discussed with respect to monitoring).

5.2.5 A note on terminology

A variety of terminology is used in the design, implementation, monitoring, and evaluation of projects. However, the basic concepts and phases tend to be similar regardless of the terminology. In Chapter 3, we suggested that when a project approach to fisheries management is desired, Logical Framework Analysis can provide a well-documented, useful tool. LFA is similar to the approach outlined in this chapter. In LFA, the overarching objectives are referred to as the "project purposes." The intermediate objectives and sub-objectives are referred to as "outputs," impact variables are referred to as "objectively verifiable indicators," and context variables are dealt with by identifying them and making assumptions about how they will change during the project implementation period. Different donors and agencies require different project formats; therefore, the small-scale fishery manager should become familiar with the requirements of donors most active in their area.

5.3 Information categories

The preceding chapters identified variables essential to the fishery management project cycle. There is also a growing literature (for example, World Bank 1999; Pomeroy *et al.* 1997, 1999; Novaczek and Harkes 1998) concerning factors influencing success of fisheries comanagement and community-based management, two approaches advocated in previous chapters. This section of the chapter organizes these variables into categories to facilitate the process of fishery management project assessment, monitoring, and evaluation, providing detailed definitions where appropriate.

For purposes of this section, the variables are divided into the following categories: 1) project variables, 2) context variables (supra-community and community), and 3) impact variables (intermediate and ultimate). These categories will result in a reorganization of some of the variables described elsewhere in this book. For example, Chapter 8 includes a list of 17 key conditions for fishery comanagement. Some of these refer to pre-project aspects of the community (for example, aspects of group cohesion such as degree of community homogeneity), while others refer to project objectives, strategies, or tactics (for example, a strong comanagement institution). In project monitoring and evaluation, the former are external factors that affect project activities or contextual variables, whereas the latter are project effects, as described above. For analytical reasons these are kept separate in this chapter, since the contextual variables are assumed to affect the project variables. The Appendix details methods for obtaining the information described in the following sections.

5.3.1 Project variables

Much has been written on the potential effects of project implementation procedures on project success in general (for example, Chambers 1983; Cernea 1991) and fisheries and coastal management projects in particular (for example, Pomeroy 1994b; this volume, Chapter 8). Early and continuous participation of potential beneficiaries is fundamental to achievement of project success and sustainability (cf. Cernea 1987; Pomeroy *et al.* 1997; Chambers 1983; White *et al.* 1994a,b). Early involvement of the community can make project objectives congruent with participants' needs, a factor linked to success of fisheries and coastal management projects (White *et al.* 1994b; Pomeroy 1994a). If beneficiary participation is real and continuous throughout project implementation, projects will need to adapt to frequent inputs from participants; hence, adaptive flexibility must be a component of project implementation (cf. Uphoff 1991). What better way to fulfil Ostrom's design principles of sustainable community-governed commons that call for "congruence between appropriation and provision rules and local conditions" and "collective choice arrangements" (1994:37) than to have local participation in the development and implementation of the rules? Participation also reduces the potential impact of two of Ostrom's "threats to sustainable community development": "blueprint thinking" (for example, universal project models) and "international aid that ignores indigenous knowledge and institutions" (1994:42, 45). The benefits of participation, known to international development social scientists for decades, have only recently been acknowledged in coastal resource management.

After an in-depth review of fisheries and coastal management projects dealing with coral reefs, White *et al.* (1994b) conclude that it is extremely important to identify a core group in the community to be trained for leadership in resource management. They also emphasize the importance of community education and training, as well as the need to communicate clearly defined project objectives to the participants. Another important aspect of project implementation is communication and coordination between groups involved with the project (for example, NGOs, government agencies, research institutions, aid agencies) (see Agbayani and Siar 1994). Fisheries and coastal management projects are frequently characterized by multiple agency and multiple group involvement, which require careful coordination. Several other researchers dealing with fisheries and coastal management projects have indicated the importance of full-time development workers and community organizers living in beneficiary communities (cf. Alcala and Vande Vusse 1994; Agbayani and Siar 1994). Lastly, sufficient, timely, and sustainable funding is critical to success of any project.

All these considerations result in the following list of indicators for evaluation of project implementation: 1) early participation in project planning, 2) continued participation in planning and implementation, 3) flexibility to adapt as project is implemented, 4) full-time development workers and community organizers living in project communities, 5) identification of a core group of participants for leadership development, 6) establishment of community education associated with project objectives, 7) coordination of all involved groups, 8) communication of clearly defined objectives to participants, and 9) adequacy of financial resources.

5.3.2 Context variables

Context variables are non-project variables. They form the immediate and larger context of fisheries and coastal management projects (for example, the local community and physical environment, the larger sociopolitical matrix). Research has indicated that these contextual variables can play an important role in development projects in general and fisheries and coastal management projects in particular. Variables identified in the context range from national policies to individual beliefs and aspirations. These contextual variables can be classified into three categories or levels, each of which is examined below: 1) supracommunity level, 2) community level, and 3) household and individual levels.

5.3.2.1 Supracommunity level

Foremost among the supracommunity variables related to fisheries and coastal management project success are, at the national level, enabling legislation and supportive government administrative structures. Felt (1990) suggests that successful comanagement is related to the amount of decision-making authority granted to participants. Jentoft (1989) writes that legislation delegating responsibility and authority to implement and enforce regulations is essential in enabling fishers' organizations to participate in comanagement of the resource (see also Kuperan Viswanathan and Abdullah 1994; Miller 1990). Most of Ostrom's (1994) design principles for sustainable community-governed commons depend on enabling government legislation; for example, a minimum recognition of resource users' rights to organize and develop their own institutions without interference by external government authorities. Enabling legislation is also necessary so that groups of users may be authorized to define boundaries and obtain security in tenure for resource-use rights (cf. Pomeroy 1994a; Alcala and Vande Vusse 1994), as well as so that users may participate in modifying use-right rules, monitoring observance of rules, and devising and applying sanctions for infractions (Ostrom's "clearly defined boundaries," "collective choice arrangements," "monitoring," and "graduated sanctions" design principles, respectively; 1994:37–39).

One of Ostrom's design principles, "congruence between appropriation and provision rules and local conditions" (1994:37), requires flexibility in enabling legislation. This is an important contributor to the degree of freedom that participants have to fine-tune the management options. In effect, where traditional or informal management systems exist, legislation should allow for its recognition and formalization (cf. White *et al.* 1994b). Feeny (1994) emphasizes the importance of the ability to adapt collectively agreed-upon resource-use rules to changing situations. Jentoft and Kristoffersen (1989) note that one of the important features of the successful Lofoten fishery comanagement system is its ability to adapt to local variations and its flexibility in response to changing conditions. The degree of adaptability obviously depends on the specificity and flexibility of government guidelines and/or directives within which the local users must work. It is unrealistic to assume that the government would delegate all responsibility for management to the resource users, with no guidelines whatsoever.

Fisheries and coastal management projects can also be supported or thwarted by supracommunity-level organizations or institutions, both governmental and non-governmental (cf. Pollnac 1994). For example, local fishers' cooperatives may belong to a regional fishers' cooperative organization, which may, in turn, belong to a national organization. Usually this hierarchy of organizations provides supportive services, such as linking the local organization to higher level sources of support. Ostrom refers to this hierarchical type of organizational structure as "nested enterprises," another of the design principles of sustainable community-governed commons (1994:41). Sometimes the supracommunity institutions are not directly related, as with the fishers' cooperative example, but are in a position to perform services (for example, training, information, supply, marketing, conflict resolution) for local-level fisheries and coastal management project activities (White *et al.* 1994b). The government may delegate authority for aspects of resource management to the local community, as in the Philippines, but lacking certain types of support and changes (for example, political will, funding for surveillance and enforcement activities, restructuring of administrative and institutional arrangements), local measures may prove to be ineffective (cf. Pollnac and Gorospe 1998; Pomeroy and Pido 1995). Pinkerton (1989b) sees the existence of institutions that provide a higher authority for appeal in terms of local equity, or institutions providing external support and forums for discussion (for example, university scientists), as favourable preconditions for comanagement.

Other supracommunity variables that can affect the success of fisheries and coastal management projects are aspects of regional, national, and international markets, including the potential for changes in commercialization of resource products. Issues such as demand and price can affect resource use and rule compliance (cf. Pollnac 1984; Pomeroy 1995). Ostrom (1994, pp. 43–44) considers "rapid exogenous changes" as a threat to sustainable community-governed commons. The changes that most threaten coastal resources are the market-related variables discussed above and new technologies. Externally developed technical changes, which diffuse rapidly throughout a fishery, can affect community-based coastal resource management systems in a number of ways (cf. Akimichi 1984; Matsuda and Kaneda 1984; Ohtsuka and Kuchikura 1984; Pollnac 1984; Miller 1989). For example, some changes might result in more efficient fishing technologies that could lessen the effectiveness of temporally based resource management regulations (for example, open seasons). Others, such as the development of more seaworthy, mechanized vessels, could result in "outsiders" long-distance fishing in local waters. Other rapid exogenous changes to consider are political instability, which could influence enabling legislation and be a variable antecedent to market instability. Natural or man-made disasters (for example, earthquake, floods, war) can imperil fisheries and coastal management projects (World Bank 1999); these should be identified, if possible. In the Logical Framework Analysis process, these are these variables about which assumptions must be made and stated.

5.3.2.2 Community level

Many of the variables that directly affect the success of fisheries and coastal management projects are found at the community level. At this level, we concentrate on both the physical and social environment in terms of potential relationships with fisheries and coastal management project success. A review of the literature, as well as information included in other chapters of this book, indicates that a number of community-level, contextual factors are proposed as being related to the success of fisheries and coastal management projects: 1) crisis in resource depletion perceived by local leaders (Pinkerton 1989b), 2) target species composition, distribution, and importance (Mahon 1997; Pollnac 1984), 3) environmental features influencing boundary definition (Pollnac 1984), 4) technology used to extract coastal resource (Mahon 1997; Akimichi 1984; Matsuda and Kaneda 1984; Ohtsuka and Kuchikura 1984; Pollnac 1984, 1994; Miller 1989), 5) level of community development (Schwartz 1986; Pollnac 1988; Poggie and Pollnac 1991b), 6) degree of socioeconomic and cultural homogeneity (Jentoft 1989; Pinkerton 1989b; Doulman 1993; White *et al.* 1994a), 7) tradition of cooperation and collective action (Pomeroy *et al.* 1997; Jentoft 1989), 8) population and population changes (Novaczek and Harkes 1998; McGoodwin 1994; Pollnac 1994), 9) degree of integration into economic and political system (Doulman 1993), 10) occupation structure and degree of commercialization and dependence on coastal resources (Pollnac 1984, 1994), 11) local political organization (Pinkerton 1989b; Pollnac and Sihombing 1996), 12) supportive local leadership (White, *et al.* 1994b), 13) quality of local leadership (World Bank 1999), 14) coastal resource use rights and management systems, formal and informal (Pinkerton 1994, 1989b; Pollnac 1994; Pomeroy 1994b; White *et al.* 1994a), and 15) local resource knowledge (see Chapter 8). The structure and content of the above variables are detailed in the Appendix. Relationships of the variables with aspects of fisheries and coastal management projects are described in the references cited.

5.3.2.3 Individual and Household Level

In the final analysis, the individual is responsible for making the decision to carry out fisheries and coastal management project (F&CMP) activities. Numerous researchers have indicated that individual variables may influence receptivity to F&CMP. For example, Feeny (1994:26), based on a review of the literature, suggests that individual variables such as education, experience, size and scope of operation, technology, "cultural values," degree of lifetime commitment to the industry, and "their preferences over non-pecuniary aspects of their employment" all have some impact on the effects of resource management regulations. The latter two of these variables have been referred to as "job satisfaction" (cf. Pollnac and Poggie 1988).

Users' ecological knowledge is increasingly recognized as both influencing receptivity to and providing information for governance (cf. Wilson *et al.* 1994; White *et al.* 1994b; Ruddle 1994b; Felt 1994; Chambers 1983; Johannes 1981), use rights, and actual management efforts. This ecological knowledge includes perceptions of the status of the resource, a variable which Pinkerton (1989b) related to development of comanagement systems. It is important to note that ecological knowledge varies intraculturally (cf. Felt 1994; Berlin 1992; Pollnac 1998, 1974). Some of this variation is related to division of labour in the community (for example, if females glean for shellfish, they will know more about shellfish than males), and some is related to degree of expertise, area of residence, fishing experience, and other factors. Hence, if we want to understand factors influencing success of fisheries and coastal management projects, it is essential to understand variation in user ecological knowledge.

In summary, the individual-level variables that may influence fisheries and coastal management projects are: 1) education, 2) experience, 3) size and scope of operation, 4) technology used, 5) "cultural values," 6) job satisfaction, 7) ecological knowledge, and 8) occupational multiplicity.

5.3.3 Impact variables

The impact variables comprise two subsets: first, achievement of intermediate objectives, and, second, effects on the well-being of the fishery ecosystem. The first subset includes consideration of degree of achievement of both material (for example, change in mesh size, organization building) and nonmaterial (for example, training, institution building) objectives. The second subset of dependent variables, the ultimate evaluation of project impact, includes consideration of project influences on the well-being of the coastal ecosystem, which is composed of separate measures of the human and natural components. Each of these components is composed of distinct sets of variables (for example, changes in income, access to resources, availability of resources). Project sustainability and cost versus benefits are also components of project success. Although categorized as dependent variables for one level of analysis, the achievement of intermediate objectives can also be conceptualized as antecedent to project impacts on ecosystem well-being; hence, as independent variables for the final level of analysis.

5.3.3.1 Intermediate impact variables

Evaluation of achievement of intermediate objectives, basic to any project evaluation, is a matter of determining whether the objectives stated in project planning documents have been met. For example, one objective might have been to set aside 50 ha of inshore waters for a marine protected area. Was this accomplished? Another objective might have been to obtain some type of legal use-right contract for nearshore areas. Was the contract obtained? Project reports usually document the achievement of objectives. This evaluation will also determine maintenance of intermediate objectives (for example, are the artificial reefs still in existence? Is there still a functional fishers' organization in the community?), as well as achievement of institutional design characteristics associated with sustainable systems (for example, attributes associated with successful user associations or community-based management of the commons).

If the fisheries management project uses comanagement, achievement of the key conditions for successful fisheries comanagement (as outlined in Chapter 8) will be sub-objectives, hence intermediate impacts to be assessed. Although some of these (for example, enabling legislation) form part of the supracommunity context, the condition could be a project objective, hence a potential intermediate impact to be evaluated. Other conditions cited in Chapter 8, clearly part of the context that the project cannot influence (for example, qualities of local leaders, community homogeneity), are evaluated as a part of the context.

5.3.3.2 Ultimate impact variables

The second subset of dependent variables, the ultimate evaluation of project impact, includes consideration of project influences on the well-being of the coastal ecosystem, which is composed of separate measures of the human and natural components. Each of these components is composed of sets of variables (for example, changes in income, access to resources, availability of resources). Project sustainability and cost versus benefits can also be considered as components of project success.

5.4 Preliminary Assessment, Baseline, Monitoring, and Evaluation Methods

Here we look at the approach and methods used in the preliminary assessment, baseline, monitoring, and evaluation of a fishery management project. This overview is more extensive than that for the other steps because it is concerned with only a subset of the information described in section 5.3 and because it requires special sampling procedures.

5.4.1 Conducting preliminary assessments

Effective fishery management project design depends on accurate and timely information on the distribution of habitats, people, and coastal activities throughout the target region. Frequently, available information is old, incomplete or unreliable. This part of the book provides a methodology to generate the information required for the early stages of project design and site selection.

5.4.1.1 Sampling for preliminary assessments

Of primary importance is adequate sampling of the diversity of communities and habitats in the region. A target region often includes hundreds of communities located in the numerous ecological niches or habitats that characterize coastlines (for example, islands, lagoons, swamps, river mouths, sandy beaches, rocky shorelines). The communities usually vary in their emphasis on productive activities (for example, farming, fishing, industry, tourism) and in their activities within these categories. For example, those characterized as "fishing communities" usually vary in the species they target and the methods they use. Faced with all this variability and the need to select a few communities for pilot projects, it is necessary to somehow describe the range of variation and select communities that represent points within this range. This maximizes the likelihood that the lessons learned in the pilot communities are applicable to the widest range of communities in the target region.

The accuracy of the description of the range of variation in the target region depends largely on the communities selected to provide data for the analysis. This choice, being a sampling problem, is temporally and financially constrained. Ideally, if the sampling universe were large enough (for example, hundreds or thousands of communities) and the budget and time constraints were generous, simple random sampling could be used, with the exact sample size being based on some type of statistical power analysis (see Cohen 1988). Also ideally, if the sampling universe were small enough (less than 30) and the budget and time allowance large enough, all the communities could be surveyed. However, time and budget constraints often rule out even stratified random sampling, resulting in the need to use some form of purposive, representative sampling to achieve a minimally acceptable profile of the target region. Since many good books have been written on the topics of simple random and stratified random sampling techniques (for example, Hedayat and Sinha 1991; Henry 1990; Rosander 1977), these will not be covered here. However, we do describe the purposive, representative technique.

Purposive, representative sampling is a technique used when financial and temporal constraints prohibit a statistically acceptable sampling procedure. Variables used to make the sample representative are similar to those that would be used to stratify a simple random sample. As long as its limitations are understood, it is the minimally acceptable method for characterizing the variation in a region of coastal communities for purposes of selecting "representative" sites for pilot fishery management projects. The limitations are as follows:

- Results cannot be used to estimate population (for example, regional) parameters. For example, if 20 percent of the sample sites manifest a certain characteristic, we cannot claim that 20 percent (with error estimates) of the communities in the region manifest this characteristic.
- The smaller the sample, the more likely it is that significant variation in the sampling universe will be missed.

Given these caveats, the purposive, representative sampling procedure should begin with a determination of maximum possible sample size, as determined by available time and funds in light of the average time required to collect data and travel between sites. Ideally, this will result in a possible sample size between 20 and 40. This should be sufficient to provide a minimally acceptable characterization of the coastal communities in the target region.

The steps in selecting a sample of villages to be surveyed for a rapid appraisal of a region are summarized below:

1. Determine time and financial constraints.
2. Using people who know the area, and available maps (preferably recent, detailed topographic and aerial charts), estimate travel time for various distances between coastal communities, using various available means of transportation (for example, boat, motorcycle, automobile, bus).

3. With knowledge of available time and estimated travel time, and assuming on-site data collection time to be a minimum of 24 hours, calculate maximum sample size and subtract 20 percent to allow for unexpected problems (for example, engine failure, severe storms).
4. Compile available secondary information on coastal communities and areas. This should include reports and statistics from regional and national statistics offices; fishery, agriculture and forestry offices; and the most recent detailed topographic, bathymetric, and aerial charts, if available. Collect legislation applicable to the area, since it may indicate sanctuaries, closed areas, and so on. Interview available knowledgeable local experts (for example, university researchers, fishery agents, city-based business people who conduct transactions in the coastal communities).
5. Examine available information and select criteria for sample selection based on what is available (for example, population, percent fishers, fishing gear types, geographic distribution, coastal characteristics: percent mangrove cover, presence of coral reefs, river mouth, island or mainland, rocky or sandy coastline).
6. Select sites based on these criteria.

If the above procedure indicates a sample size less than 20, either adjust available resources to increase sample size or accept the fact that the limitations noted above will apply more severely and reduce the reliability of the assessment to a level that may be unacceptable.

5.4.1.2 Personnel for conducting preliminary assessments

Since the amount of time allocated for each community during such a survey will be minimal (that is, one to two days), it is important to specify the desired characteristics of field workers, preliminary preparations necessary, transportation and accommodations, and limitations of the data.

The field team should be small (no more than three or four, including boatman or driver) to facilitate movement and accommodations, as well as to minimize the disruption in small communities. At least one member of the team should be fluent in the local dialect. The scientists (social and/or biological) should have broad experience in coastal communities, with extensive knowledge of traditional and modern coastal resource productive activities; for example, they should be able to recognize most fishing gears and identify organisms with the aid of a guidebook. Team members should be in good physical condition, able to walk tens of kilometres a day, day after day, in the prevailing local weather conditions. They should be sufficiently adaptable to go to sea with local fishers to observe fishing techniques if necessary, and at least one should be able to use snorkel gear to observe general condition of aquatic habitats. Ideally, one researcher would be a social scientist with extensive experience in small-scale and industrial fishing communities and the other a marine biologist.

5.4.1.3 Information for preliminary assessments

Since the information requirements for the preliminary assessment are less than for the baseline, monitoring, and evaluation process, the following outline is brief. Because inadequate time is usually allocated for the preliminary appraisal, it is important to specify a minimal data set that will provide a sufficient though superficial sketch of fishery activities and conditions in the target area. Hence, every variable specified requires a clear rationale.

Population amount, distribution, and density are clearly related to fishery ecosystem health, influencing both pollution and intensity of resource exploitation. Productive activities (for example, fishing, farming, coral mining, mangrove cutting) are directly related to fishery ecosystem health, and the occupations associated with them are an important aspect of community social organization. Discovering the existence of potentially destructive practices associated with productive activities is especially important. Different relationships between the people and the natural resource are often reflected in and influenced by community social groupings (for example, organizations, ethnic, and religious groups); hence, the social groupings must be accounted for in a preliminary appraisal for fishery management purposes. Community infrastructure (for example, roads, schools, medical care, markets, transportation) is linked to many aspects of the fishery ecosystem, especially the economic value of fishery products and the quality of life of the human population.

It is also important to determine issues such as perceived changes over the past five years in overall well-being of the community, condition of the fishery, and condition of other coastal resources exploited. Reports on these issues from key members of the community provide information that otherwise may be impossible to discern in a brief visit. Finally, a general description of the coastal geography (for example, outstanding oceanographic conditions, such as destructive currents or wave action; ocean depth near shore; minimal description of coral reefs; distribution of mangrove; beach characteristics, including litter and erosion; locations of rivers, streams, and swamps) is necessary. Information on coastal geography will facilitate understanding of existing relationships between the local population and their fishery environment, as well as indicate potential problem areas. The foregoing represents the minimum, essential data needed for an initial understanding of fishery management issues for a target area.

Following are more detailed descriptions of the variables to include in a preliminary assessment:

1. **Coastal zone physical geography:** general description of terrestrial terrain (for example, slope, land use); general coastal configuration and condition (for example, bay, river, swamp, and mangrove locations, estimates of size, shapes), composition (for example, sand, pebbles, rocks) and extent of erosion, litter, and runoff; nearshore bottom characteristics (slope, presence and general condition of coral); and any salient climatic or oceanographic conditions that influence human behaviour (for example, strong currents, large waves, seasonal storms).

2. **Population:** present population of the community and population from the next previous census, in order to evaluate recent population trends.
3. **Settlement pattern:** are the houses and other structures concentrated in one area (for example, along the coast) or are they dispersed, or in some combination of settlement patterns (for example, nucleated [concentrated] on the coast and dispersed inland)?
4. **Land:** area and suitability for agriculture.
5. **Occupations:** percent of population engaged in various occupations.
6. **Fishery activities**: for each activity, identify target resource, methods, and gears (numbers of gears), who is involved, when, where, why (home consumption, market), approximate catch effort information, and method of marketing and distribution.
7. **Community infrastructure:** number of hospitals, medical clinics, resident doctors, resident dentists, secondary schools, primary schools, water piped to homes, sewer pipes or canals, sewage treatment facilities, septic/settling tanks, electric service hook-ups, telephones, food markets, hotels or inns, restaurants, gas stations, banks, public transportation, and paved roads.
8. **Social groups:** percent distribution of ethnic and religious groups; names of all organizations, identified by type, function, year formed and membership.
9. **Major issues:** perceived changes over the past five years in: 1) overall well-being of the community, 2) condition of the fishery, and 3) condition of other coastal resources exploited.
10. **Destructive practices:** presence of destructive techniques such as poisons, dynamite, or scare lines over coral reefs; anchoring on reefs; pollution of waters; and so on.

5.4.1.4 The field work for preliminary assessments

All secondary information available should be reviewed and the required data abstracted for communities in the sample. Charts should be carefully scrutinized and preliminary travel plans developed, allowing for flexibility, since field conditions may be better or worse than those depicted on the charts. If available, local terminology for coastal resources (flora, fauna, mineral, etc.) and gears and techniques should be compiled to facilitate data acquisition. If local taxonomies are not available, extra time should be allocated for the first site, or a trip should be made to a community in the region to compile preliminary taxonomies. A preliminary taxonomy should include most important species and gears. This can be supplemented as the data collection procedure goes forward.

If necessary, permission for travel through the area should be obtained. Also, if necessary, letters explaining the purpose of the exercise should be sent to local community leaders to prepare them for the team's arrival.

Such an assessment is best conducted by boat. Access to marine sites is facilitated, poor coastal road conditions (or lack of road) are irrelevant, and if the boat is large enough, the accommodations problem is avoided. One of this book's authors conducted a similar assessment using a 9- by 2-metre boat with a small open cabin. The cabin was extended with a wooden framework and a tarp, and the team (human ecologist, marine biologist, boat driver, and helper) slept on the boat, just offshore at the communities in the sample; thus, they could observe coastal activities round the clock.

If transportation is by land, planning should account for the fact that many coastal communities are difficult to access. The map may show a road connected to the coastal community, but roads along the coast are frequently in poor condition, and the community centre may be several extremely rough kilometres away from the coastline and coastal residents. Sometimes the coast cannot be accessed by motor vehicle, necessitating a time-consuming hike through terrain containing little information of use for the preliminary assessment.

If one must arrange accommodation, all attempts should be made to stay in the sample community, despite the fact that most small rural coastal communities do not have hotels or inns. Time is of essence in this type of survey, and time spent traveling back and forth to an inn or hotel in another community is wasted. It is usually possible to find someone in the sample community who has a spare room, but the team should be prepared to sleep on the floor. Accommodation in the sample community provides extra time, while eating and settling in for the night, to acquire information that might have otherwise been missed.

It is important to note that the limited time spent in each village places constraints on the process of validating information acquired by interview. In many cases, it is possible to make observations that can be used to validate certain types of information. For example, if told that a certain type of resource use is carried out at night, attempts should be made to observe the practice; if told that no mangrove were harvested recently, the mangrove area should be examined. Observed numbers of boats, by type, can be used to validate statements about approximate numbers of fishers. Some productive equipment, however, is small enough to be kept in the household or other closed storage place; hence, it is necessary to rely on informants' information (for example, local fishers, fish buyers).

One problem with making observations (and conducting interviews) concerning natural resource use is that the activity is periodic (for example, fishing seasons or times) and often conducted in difficult-to-access areas (for example, on the far side of an offshore island). Use of a boat facilitates assessment of a wide range of areas not readily assessable from land. Nevertheless, the periodicity of activities can influence what informants say (for example, they are more likely to respond with information concerning current activities), as well as the scope of observations, since interviews never provide the insights gained by observation of the activity. This is a great weakness of rapid assessment techniques, especially with respect to fishery activities.

Nevertheless, with these caveats in mind, information derived from interviews with several informants, as well as observation where possible, does present a relatively reliable "snapshot" of conditions in the sample communities and practices relevant to fishery management. Such information should, however, be used only as a preliminary overview, to stimulate further investigation to derive information on which to base coastal management efforts.

Table 5.1 provides a summary of preliminary appraisal data needs cross-tabulated with methods discussed in section 5.4.2. This table should be used with the text, since it is a superficial summary.

Table 5.1 Cross-tabulation of data-gathering techniques and variables

Variable	Secondary Data[A]	Community Officials[B]	Key Informants[C]	Observation[D]
Population	X	X		
Settlement pattern	X	X		X
Land area	X	X		
Arable land area	X	X		
Occupations	X	X	X	X
Fishery activities	X	X	X	X
Community infrastructure	X	X	X	X
Social groups	X	X	X	
Major issues		X	X	
Destructive practices		X	X	X
Illegal practices	X	X	X	X
Coastal geography	X			X

A Secondary data include published statistics, reports, maps, legislation, etc.

B Community officials include mayor, chief, secretary, etc.

C Key informants refers to any knowledgable persons, including those inside and outside the community, such as government agency personnel who have visited the community, researchers who have worked in the area, and community members involved in the activity being investigated.

D Observation refers to observations made by the research team during beach walks, the community walkthrough, sailing by on a boat, while participating in activities,and at all times while in or near the community. It should be a constant activity.

5.4.1.5 Conducting a baseline survey

Once pilot sites for fishery management have been selected on the basis of the preliminary appraisal, it is necessary to obtain more detailed, baseline information from communities selected for pilot projects. The baseline must include information on the context and impact variables discussed in section 5.3. Concurrent with the collection of baseline information at the pilot project sites, it is necessary to collect the same type of information from nearby communities to be used as control sites. The baseline includes many of the same types of information that are included in the preliminary assessment, as well as additional information. The main difference is the level of detail, accuracy, and reliability. These differences result from the use of different methods (survey), including the expenditure of a greater amount of time in data collection. For example, while little more than 24 hours may be spent in each community for the preliminary appraisal, it may take several weeks to collect and analyze the baseline data for one community.

5.4.1.6 Conducting monitoring and intermediate impact evaluation

Ideally, the monitoring process begins upon implementation, as soon as project activities directed at achievement of intermediate and ultimate objectives begin. It is not realistic to expect measurable changes with respect to ultimate objectives in the first few years of a project. One can, however, assess project activities and changes in contextual variables, identify issues missed in baseline assessments, and evaluate achievement of intermediate objectives. A common strategy for both implementation and monitoring of fishery management projects is to assign extension workers to live in the target communities. They identify issues that may have been missed in initial baselines and help to better understand their social and political context. The extension workers obtain this information through such techniques as community immersion and participant observation; long-term direct observations; informal individual, key informant, and small-group discussions; formal focus groups or community meetings; and community mapping, as well as other participatory and non-participatory appraisal methods (see IIRR 1998).

Another type of required monitoring concerns the achievement of sub-objectives. These usually constitute the strategies implemented to achieve the ultimate project goal. Typical strategies for improving the health of the fishery often include cessation of destructive fishing methods and establishment of MPAs (for example, marine sanctuaries, protected areas). The sub-objectives themselves frequently involve strategies that can be conceptualized as a series of sub-sub-objectives. For example, for a fishery management project, community members must first become aware of the problem and potential solutions. This often involves a public education program that may involve community meetings, strategically placed informative posters, and so on. The public education program thus becomes a sub-sub-objective that must be monitored and evaluated. Meetings with community members must be held to select solutions from the list of alternatives. Once solutions are selected, a plan to implement the activity (for example, selection of area for MPA, identification of appropriate markers for borders, establishing surveillance techniques) must be developed.

Frequently, most of the intermediate objectives essential to achievement of a sub-objective are sequential. One must be achieved before the next one can be achieved, and before the extension team and community can move on to the next steps in the fishery management project process. In other instances, achievement of certain objectives through implementation of a set of actions will help in the achievement of other objectives. All these activities must be monitored to identify problems as they develop and adapt the strategy to achieve the objective.

5.4.1.7 Conducting ex-post evaluation

The logic of ex-post evaluation is the same as that used in monitoring and evaluation during project implementation but is more extensive. Several data sets are developed to conduct the ex-post evaluation. First, the variables included in the baseline for project and control communities (both human and natural environment) are collected again. This information is used to make time-one, time-two comparisons of human and non-human aspects of the ecosystem to assess fishery management project impacts on ecosystem health — the ultimate project objective. This data set also includes non-project, contextual variables that may help evaluate alternative explanations for observed changes. Second, all interim monitoring and evaluation reports are collected. Information in these reports can be used to identify both project and contextual variables that may account for observed changes. Third, the status of all sub-objectives must be assessed. Ideally, the monitoring program reports will help in the formation of this data set. It is, however, essential to have current information on the status of interventions such as MPAs, gear restrictions, and so on. Some of these activities may have ceased or degraded since implementation. A fourth data set includes "shocks" (for example, changes in markets, a new road, typhoons, wars) to the system. Finally, the fifth data set includes villagers' perceptions of changes in human and natural components of the fishery that the fishery management project is meant to improve. Although the other data sets allow us to assess these changes, it is the villagers' perceptions of project impacts that influence their behaviour in ways that can ensure project sustainability.

5.4.2 How do we get the information? — Data collection techniques

The collaborative approaches to fishery management that this book advocates also include data collection. While some types of information must be acquired, or at least verified by non-community members to ensure objectivity, some can and should involve community members. This type of collaboration gives community members a greater understanding of and a vested interest in the management process. Additionally, the community perspective is more likely to result in a successful project. Methods for involving coastal community members in data collection have been developed and are clearly described in other manuals (for example, Walters *et al.* 1998; IIRR 1998).

Box 5.6 Community involvement in data collection.

As part of a coastal management project in the village of Blongko, North Sulawesi, Indonesia, community members were trained to use a manta-tow method to monitor changes in coral reef health. Data collected by an independent professional survey was compared with data collected by the community. A matched-pair T-test indicated no statistically significant differences between the two data sets ($p > 0.05$). This finding supports the use of properly trained community members in project monitoring and evaluation (Fraser et al. 1998).

With or without collaboration, four principal data collecting techniques are used for collecting the information types operationally defined below: use of secondary information, observation, key informant interviews, and sample surveys. We briefly describe each, mentioning its benefits and disadvantages. For more detailed information on these methods, see Schensul and LeCompte (1999) or Pelto and Pelto (1978).

5.4.2.1 Secondary information

Secondary sources include official and unofficial documents, statistics, and maps. These can be important for obtaining preliminary information as well as that to be used to cross-validate information gathered by other techniques. One of the benefits of using secondary information is that it prevents duplication of effort — much time and effort can be saved if reliable information is already available. Disadvantages are that reliability and validity of information may be difficult to ascertain, and information is frequently out of date (see examples in the data quality control section below). Evaluation of documents requires special attention. A great deal of evidence indicates that questionable, if not outright erroneous, findings are published in refereed journals and books, as well as in the "grey literature" that serves as an outlet for much applied research (Katzer *et al.* 1982). It is important to note that much of the information on coastal zones in developing countries is published in grey literature. Over 30 years ago, Naroll (1962), in a book on data quality control for quantitative cross-cultural research, argued that the researcher has a duty to evaluate report reliability. For more information on data quality control in the use of secondary literature, see Pollnac (1998), as well as the other references cited above.

5.4.2.2 Observation

Observation, in which the investigator attentively watches and records events, is one of the most useful field research techniques because it is a direct means of collecting information on activities, including the roles of various participants. Observation is especially important for deriving information from a fishery because much of the behaviour involved in these activities is carried out using tacit knowledge, learned nonverbally, by observing and doing. For example, fishers find it difficult to verbally describe all that they do while at sea. Consequently, significant information may be missed if a researcher relies only on the interview process. It is thus necessary to observe, or even participate in, the activity to acquire a full understanding of all its aspects. Observation can also lead to new insights and discoveries, explain activities that are difficult for participants to describe, help in the formulation of interview and survey questions, and verify information derived from other sources, such as secondary data or interviews.

5.4.2.3 Key informant interviews

Key informant interviewing requires identification of individuals with knowledge that can contribute to the sought-after information. For example, the head (mayor, village chief, etc.) of a village or town can provide much information concerning his area. Likewise, fishers know much about their fishery. The advantage of using key informants is that a researcher can rapidly obtain information on a topic without observing it or conducting a survey. The disadvantages are that a key informant may possess inaccurate information or purposefully provide misleading information (see examples in data quality control section below). For these reasons is it essential to cross-check key informant information with several other key informants and to observe.

Key informants may be interviewed individually or in groups. Group interviews (that is, focus groups) should be used with care, since some individuals may dominate the discussion or, because of their status, inhibit refutation of their viewpoints. Sometimes group interviews tell you more about group interaction than the topic being discussed. For these reasons, all information from groups should be cross-checked through observations, interviews with several other groups, and/or individual key informants.

Furthermore, the use of focus groups, being a formal technique, is time-consuming. Individuals must change their schedules to attend the focus group meeting, an imposition that may negatively impact responses, possibly reducing the quality of data obtained.

Key informants, individually or in groups, can also be used in more participatory approaches. For example, fishers can help to construct maps illustrating distribution of resources and fishing areas. Especially knowledgeable and enthusiastic individuals sometimes become part of the information gathering team, further enhancing the level of community participation.

5.4.2.4 Sample survey

Sample surveys involve the development of an interview that is administered to a sample of respondents in the target population. They are very important because they can result in the quantification of variables and can be used to cross-validate information from other sources. The principle strength of the sample survey is that results better represent the population being investigated than does information from key informants or focus groups. Additionally, results are amenable to statistical analyses. These strengths more than outweigh the disadvantage of higher time and money costs. Sample surveys, however, lack the in-depth, qualitative information that can be obtained through observation and key informant interviews; hence, they should be used in combination with these other techniques.

5.4.3 What do the numbers mean? — Data quality control

Data quality control is always an important issue in fishery management. It becomes more important as arguments are made for using less information in terms of smaller samples (if any), lower levels of measurement (for example, nominal or ordinal as opposed to interval data), and information on a reduced number of variables. Whereas it is relatively easy to say that suboptimal management of fisheries is better than no management, there are doubtless examples to the contrary. And we have to ask, better for whom? Humans are part of the ecosystem managed by fishery managers. If inadequate or inaccurate information results in restrictions that reduce fishers' income or food supply, especially in the many cases where the people have no alternatives, is this better? Can we condone a situation where a man cannot pay school fees for his child or, even worse, cannot put food on the table because of "suboptimal" management? And if this suboptimal management is found at some later date to be based on faulty data, who will convince the fishers that subsequent, perhaps more appropriate, management plans should be followed? Hence, in any human-centred approach, we should be deeply concerned with the reliability and validity of the information used in management decision-making. This concern involves three important, interrelated issues: validity, accuracy, and reliability of information. Each of these concepts is explained below (see also Chapter 6 for sources of uncertainty in fisheries management).

5.4.3.1 Validity

The question of validity asks whether a given measure constitutes a correct measure of the phenomena under consideration. Measure, as used here, can refer to an interval measure of a quantity (for example, tonnes of shark), ordinal measure (for example, more shark than last year), or nominal measure (either shark or no shark). An example of a question of validity is whether it is correct to base a statement of ordinal quantity on the response provided by a respondent (for example, fisher, fish dealer, etc.) when asked if there are more shark today than five years ago. Here the issue is whether the response reflects reality, a perception of reality, or a desire to provide a response that will impress or influence the questioner in some manner. As long as we qualify the quantity as "perceived" quantity, the method is valid. But any attempt to use perceived quantity as a real quantity in a management plan would be invalid, since perceptions can be based on faulty assumptions or incomplete data. For example, fishers from some societies may say that there are the same number of fish even though catches are declining because they believe that spirits are keeping fish away from the fishing gear (see Zerner 1994). Others may base their responses on observations of their own catches that may differ significantly from others' observations. Hence, the technique would be a valid measure of perceived ordinal quantity that may or may not reflect the actual ordinal quantity.

5.4.3.2 Accuracy

The question of accuracy is concerned with the degree of precision with which a given technique can measure the variable of interest. A great deal of research has demonstrated that informant accuracy in recall differs significantly from information based on direct observation of the phenomena to be recalled (see Bernard *et al.* 1984 for a review of the literature); in one case where it was quantified, accuracy varied by as much as 56 percent (Ricci *et al.* 1995).

5.4.3.3 Reliability

A reliable technique will result in the same measure each time it is applied unless the variable being measured has actually changed in value. With regard to many of the variables examined in this chapter, researchers frequently depend on verbal interviews with community members. If this is the case, we can question the reliability of claiming that the responses of several individuals can be generalized to the entire community or local fishery. We need to ask, if we interview several more individuals, selected using the same criteria, would we obtain the same information? This, of course, depends on data type (see Poggie 1972), but, in general, generalizing from a few interviews to the larger population can be quite unreliable. Some have suggested that community meetings or focus groups can eliminate this problem, but recent findings indicate that such techniques can also be unreliable. For example, Davis and Whittington (1998) have shown gross and statistically significant differences between results obtained from community meetings and household surveys regarding variables such as percent of households with certain services, percent home ownership, and willingness to pay for certain services. In general, focus groups and community meetings reveal much about group dynamics but provide limited reliable information.

The following boxes illustrate several examples of potential problems in data collection that show the need for concern with quality control of data. The first example involves an apparently straightforward and important variable — the number of fishing vessels. The second, like the first, demonstrates the need to cross-check information sources, but with a different data type — apparent conservation practices.

Box 5.7 Quality problems with secondary information.

Prior to conducting field research in a coastal area in the Philippines, secondary material was reviewed to obtain preliminary information concerning the fishery. The secondary information was only two years old, so it was expected to be relatively accurate. Researchers boating into a community, which had been reported to have three motorized and four non-motorized boats, noticed at least 30 boats moored along the beach. In the community, the ex-village chief, who is an active fisher, living among the fishers, reported 50 unmotorized and 3 motorized boats. The village secretary stated that there were 150 non-motorized and 10 motorized boats. The researchers only had a brief time allocated to obtain this data in the community, so after leaving they obtained figures from several other sources. They interviewed the official responsible for vessel registration and painting, who reported 84 non-motorized and one motorized. The office of the Municipal Agricultural Officer had 49 boats listed in his survey. This wide variation in numbers clearly shows the need to check such information.

The best method for determining number of boats is to count them at a time when most, if not all, are at the dock or on the beach. Most of the coastal villages in this example had several beaching and/or docking areas spread over a rugged coastline, making such a procedure impractical, given the time constraints of the project. The only village where this was accomplished resulted in a vessel count of 78 at 2:00 pm, when it was reported that all boats were likely to be beached. The researchers realized that some boats may have been taking fish to market, obtaining water from the mainland, or carrying out some other task. Nevertheless, the count of 78 is extremely close to the Office of the Municipal Agriculture Officer's count of 74. The Office of the Municipal Agriculture Officer list of vessel owners was checked by an Atulayan resident, who added a few names and was unsure concerning about one-fourth of the list (but could not discount ownership), resulting in a figure of 79 vessels. The closeness of this confirmation, as well as the detail in the data (the names of owners), led the researchers to select the Office of the Municipal Agriculture Officer survey as the best available information for the rest of the villages in the target area (adapted from Pollnac and Gorospe 1998).

Box 5.8 Quality problems with key informant data.

While collecting information concerning species and methods in a bay on the north coast of Jamaica, a researcher was interviewing an individual fisher that scientists at a nearby biological laboratory had recommended. They had very positive interaction with this individual, a cooperative elderly, knowledgeable fisher. When he mentioned a species caught in a beach seine, the researcher asked him why beach seines were no longer used. He said the fishers no longer used them because they knew that they took everything, small fish and shellfish, harming the resource. An interviewer with little time and the "politically correct" perception of the traditional fisher as a conservationist would have probably recorded this information and written it in a report (noting, one would hope, that it was obtained from one highly recommended fisher who had a lot of contact with the marine laboratory personnel). The investigator, a skeptic, both about the "fisher as conservationist" and the representativeness of a fisher who has had extensive contact with marine scientists and comes highly recommended, continued to hunt for other possible reasons for the end of beach seining. After a bit of probing, the fisher added that there was an economic reason. The owners of beach seines used to be "rich men" who hired labour to set and pull the net. He said that the fish caught today are so few and small, and worth so little, that fishers would no longer hire on as labour for the small amount of income they would receive; hence, the demise of the beach seine. This explanation made sense, but interviews with more fishers (ones not recommended by anyone) provided an additional, more compelling factor. Scraps of metal and cable were deposited on the bottom of the bay during the construction of the harbor for the giant vessels that haul bauxite from the local bauxite processing plant. This scrap metal snagged the beach seines in traditional seining areas, inhibiting their use; hence, another, more compelling reason for the demise of the beach seine fishery in the bay. None of the other fishers interviewed said anything that could be interpreted as a "conservation ethic," although such a response probably could have been stimulated by a question such as "I've heard that fishers quit using the beach seine because it kills the little fish and shellfish, hurting the resource. Is that true?" Investigators have actually been heard using such leading, hence misleading, questions (adapted from Pollnac 1998).

5.5 Conclusions

This chapter has outlined the general phases and logic associated with the process of assessing, monitoring and evaluating a fishery management project. It outlined the steps of the fishery management project cycle, describing the types of information needed at all stages of the project for setting of objectives, developing strategies and continually monitoring effects. Techniques for acquiring and analyzing the information were discussed, along with the importance of information reliability and validity. This information and the activities described are necessary to adapt the management process, as well as to learn from failures and successes. The processes described in this chapter facilitate this process by making sure that the manager has a clear picture of the initial stages from which the management process develops. These initial pictures are then used to determine the effects of the process when compared with comparably derived pictures of the later stages. When these pictures do not fit desired objectives, the process can be adjusted to achieve the desired ends.

Chapter 6
Fishery Management Process

6.1 Introduction

This chapter addresses the problems of making fisheries management operational. We pick up from Chapter 3, which placed a great deal of emphasis on the need for process and clear objectives for efficient management. Even with efficient planning, conventional approaches to implementing management may be too costly for small-scale fisheries. Therefore, the manager must be innovative, particularly in considering simple methods for management. Because there is no documented set of "best practices" for managers of small-scale fisheries, we present here a variety of ideas — some tried, some recently emerging, some just ideas. The manager will have to consider these in relation to the needs of the fishery in question and assemble a package that appears workable.

In tackling this difficult topic, we first develop a framework that allows the manager to move forward from the management objective to the establishment of management targets and limits, or, when these are elusive, management directions. As we emphasized in previous chapters, small-scale fisheries usually lack data, information, and a capacity for top-down enforcement. Therefore, we place a great deal of emphasis on reaching agreement among stakeholders as to what should be done.

The second part of this chapter looks at the management tools or measures available to managers for changing in the status of fisheries. Finally, we explore the problems of enforcement and compliance in small-scale fisheries.

6.2 Management process

Figure 6.1 outlines a view of the fishery management process. The sequence of activities depicted is a continuation of the process of planning and setting objectives described in Chapter 3. The emphasis in these activities is on taking the societal goals and objectives, together with the technical constraints and opportunities, and developing the operational means by which the objectives will be achieved. Typically, in conventional management, this is done by determining Target Reference Points (TRPs) on reference variables that are considered to indicate the state of the fishery. **Figure 3.3** illustrated some typical conventional and non-conventional TRPs in relation to a surplus production model. Recently, the concept of Limit Reference Points (LRPs) has come into play. To these we will add the concept of Management Reference Directions (MRD).

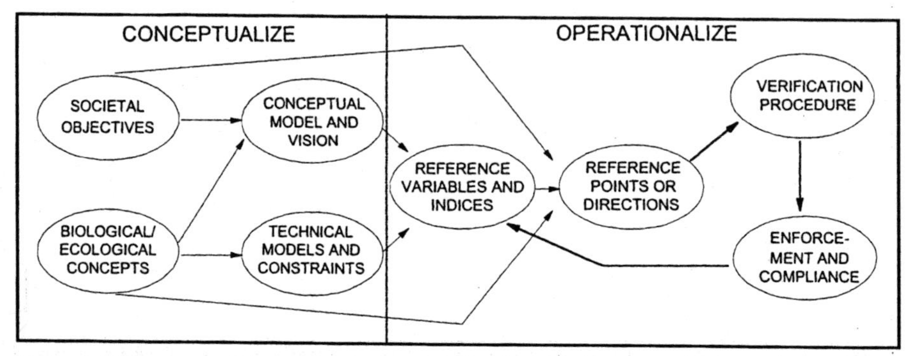

Figure 6.1 The conceptualization and operationalization of reference variables, points and directions incorporating societal goals and technical models for fisheries management.
Source: adapted from Caddy and Mahon 1995

6.2.1 Reference variables

Once the objectives of management have been agreed upon, as outlined in section 1.6 of Chapter 3, the next step is to agree upon the variables or indices that will provide the best measure of where the fishery is in relation to each objective. If the objectives have been set in terms of standard fish stock assessment models, then the reference variable will usually be fishing mortality, or fishing effort, which is often considered a reasonable proxy for fishing mortality. However, in small-scale fisheries, use of these models may not be feasible or cost-effective. Therefore, the challenge for managers of small stocks is to find and use alternative reference variables that:

- Depend less on quantitative models with high research and data collection demands;
- Better reflect the social, economic and environmental objectives.

These types of variables often take the form of indices that do not have the potential for precision inherent in the conventional models. They are more likely to represent the broad brush approach referred to in Chapter 1 rather than the narrow arrow approach of the conventional models. However, they should have the advantages of being less costly to monitor and more easily understood by stakeholders. An understanding of the relationships between indicator variables that relate directly to the fishery management objectives and the performance of the fishery is an area in urgent need of research attention. Encouraging research on these variables would be one way of improving the now-weak link between research and management (Pido 1995).

6.2.2 Reference points and reference directions

Once the reference variables have been selected, the next step is generally to select TRPs and/or LRPs on them. An LRP may be a point on the same reference variables as a TRP, or may be a point on a different reference variable. For example, a target could be a catch of 1 000 tonnes of a species; recognizing the imprecise nature of management, a limit for the same fishery could be 1 200 tonnes. Alternatively, or additionally, in the same fishery a limit could be set as not more than 200 tonnes of a bycatch of another species. Going further, if environmental impact of gear is a problem, a limit in the same fishery could be that the target catch must not be taken with more than 2 000 trap sets or trawl tows. In the latter cases when the limit is reached, fishing stops even if the target has not been reached. Caddy and Mahon (1995) provide a review of conventional fisheries management, while Caddy (1998) provides a review of precautionary reference points..

The selection and adoption of reference points is a critical stumbling block in many fishery management schemes. The emphasis on target and limit reference points is clearly appropriate when there is enough information to identify the points. When there is not enough information, Management Reference Directions may be an adequate basis for management action. This will often be the case in small-scale fisheries, particularly those showing clear signs of overexploitation. For example, in reef-fish fisheries, the conclusion that the catch consists of too high a proportion of small, low-value species plus immature individuals, provides an indication of the Management Reference Direction. The desired Management Reference Direction is to rebuild populations of valuable species, increase the size of fish in the catch, and reduce the proportion of immature individuals. The target points on these variables may not be known, but the need to move in that direction may be clear, and it may be possible to do so without knowing the target end point (**Figure 6.2**). This shifts the focus of management action from "Where do we want to be?" to "How do we move from here in the desired direction?" Generally, the latter question is easier to answer.

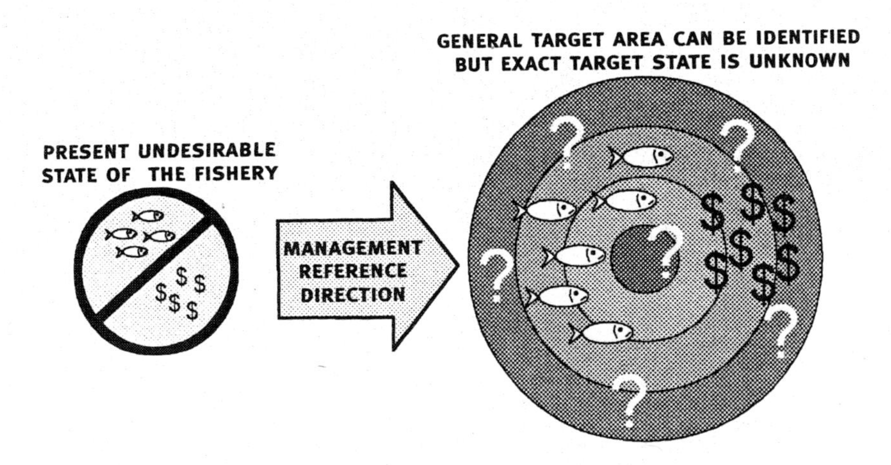

Figure 6.2 Reference directions as the basis for initiating management action even when target reference points cannot be established with certainty.

The concept of a Management Reference Direction rather than a target or limit point is a logical extension of the clauses in the *Law of the Sea*, *Code of Conduct for Responsible Fisheries* and other international instruments calling for management based on the best available scientific information. They warn that management should not be delayed while one waits for more scientific information. Incorporation of the concept of a Management Reference Direction into management planning will be a sufficient basis for action in many small-scale fisheries where problems have been qualitatively identified, but quantification is not feasible.

6.2.3 Can simple reference variables and indices be useful?

Most TRPs used in conventional fisheries management are based on biological and bio-economic models. The main reference variables for the TRPs have been: fishing effort, fishing mortality, stock biomass, spawning stock biomass, catch, revenue, and profitability. In regard to these, we ask the following questions:

1. Do any of the above reference variables provide a useful basis for reference points, even when there may be inadequate data to use the underlying models?
2. Are there more easily observed broad brush indicators that relate to the narrow arrow reference points in a way that can be useful?
3. Can any other, preferably simple, reference variables serve as useful indices of the status of the stock relative to its desired status, such as qualitative indicators?

The answer to the first question depends on whether the variable can be directly observed, or is an output of the model. For example, catches can be directly observed, fishing mortality cannot. Therefore, in the absence of data for models to estimate optimal catches or mortality, stakeholders can agree upon a catch to be used as a target but the same cannot be done for a target fishing mortality.

The answer to the second and third questions is "yes," many indices of the extent of exploitation or fishing mortality can be used. It has long been known that exploitation results in changes in individual species and also in entire communities. Recent studies have addressed these well-known phenomena and conclude that fishing produces predictable structural changes in fish communities because species respond to exploitation according to their life history characteristics (Jennings *et al.* 1997, 1999). These changes can be used as indicators of the levels of fishing, particularly when there is the potential for comparison among similar communities at different levels of exploitation. Furthermore, the changes in the communities are reflected in the catch in ways that inevitably affect the unit value of the catch. In extreme cases of overexploitation, the catch may consist largely of very small individuals and high proportions of low-value species. Thus, these changes present opportunities for establishing economically relevant targets or directions.

Following is an example, based on Mahon (in prep.) of how knowledge of species composition might be used for reef-fish fisheries. Species composition of reef-fish assemblages changes with exploitation and will relate to the various points on the curve in **Figure 3.3**. For Caribbean reef fisheries, the snappers and groupers are key indicator species. The proportion of these species in the catch varies widely from one fishery to another (**Table 6.1**). An experienced person can quickly assess the status of a reef-fish fishery by looking at the species composition, either of the catch on shore or of the of the exploited assemblages while diving or from a glass-bottom boat. Most of these persons could place a fishery or exploited assemblage on a scale 1-5: 1 = lightly exploited, 2 = moderately exploited, 3 = fully exploited, 4 = overexploited, 5 = depleted. One could argue about exact definitions of these categories, but there would be a high degree of correlation among the scores of different assessors.

There is also the possibility that, with a small research effort, an indicator of this sort could be taken to a higher level of sophistication as a possible reference variable. By taking a comparative approach to exploited coral reef–fish assemblages throughout the Caribbean region, it should be possible to develop a model relating the percentage of snappers and groupers in the catch to the status of the resource. Further refinement may be possible using the relative abundance of other species as indicators. We will illustrate this possibility with the data already available in **Table 6.1**, by assuming 1) that the relationship between species composition and effort is linear (probably incorrect), 2) that the Jamaica north coast in 1990 was at the unmanaged equilibrium for a low cost fishery with sustainable yields at about 20 percent MSY, and 3) that, due to underexploitation, Bermuda was in 1975 also at about 20 percent of MSY. With these two anchor points we can locate the other country/year data points from **Table 6.1** on the surplus production curve (**Figure 6.3**). Managers can then decide which way they want to move their fisheries and begin to discuss measures for taking them in the desired direction.

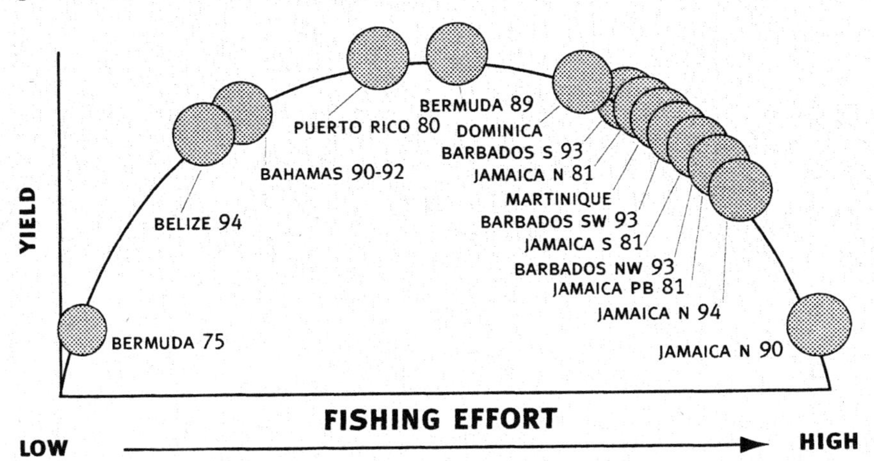

Figure 6.3 An assessment of the relative status of coral reef–fish fisheries in Caribbean countries based on the relative proportion of snappers and groupers in the catch (**Table 6.1**). Labels indicate the location and the year. These estimates are uncertain and without any confidence interval. Their exact location on the curve may vary depending on the relationship between species composition and effort.

Table 6.1 Percentage contribution of snappers and groupers to reef-fish landings in various Caribbean countries.

Place and time	Snappers	Groupers	Total
Bermuda 1975	56	11	67
Belize 1994	35	21	56
Bahamas 1990–1992	38	18	56
Puerto Rico 1980	23	22	44
Bermuda 1989	25	13	38
Dominica	17	12	29
Barbados south coast 1993	24	1	25
Jamaica North Shelf 1981	1	23	25
Martinique	12	12	24
Barbados SW coast 1993	18	3	21
Jamaica South Shelf 1981	1	19	20
Barbados 1993 NW coast	18	2	20
Jamaica Pedro Bank 1981	14	6	19
Jamaica north coast 1994	6	11	18
Jamaica north coast 1990	8	4	12

Source: Mahon, in press

Clearly, some refinement to this method is desirable and possible. However, it illustrates how a simple analysis with easily obtained data can provide some guidance to the manager deciding what to do with a fishery.

For small-scale freshwater fisheries, Welcomme (1999) observes that models which group species have proven adequate to provide the level of advice needed to indicate ecosystem health and sustainability of yield from a fishery. He suggests that these models should be further developed, noting in particular that average length of fish caught, numbers of species in the catch, and time taken for catches to respond to floods are good indicators of the health or status of the fisheries.

Another simple indicator for fisheries management is the proportion of immature individuals in the catch of a particular species. Conceptually, this corresponds to the minimum spawning biomass reference variable that has recently become popular among fisheries managers. Targets and limits on this variable have been set using sophisticated models. However, the small-scale fishery manager may be able to accomplish the same objective without the quantitative models by using a broader brush approach. For example, if 95 percent of the individuals caught of an important species are immature, most stakeholders would agree that there probably has been or will be a negative impact on the productive capacity of the stock and that there may even be a high risk of commercial extinction. It should not, therefore, be difficult to get agreement to move toward a lower proportion of immature individuals in the catch.

The next question is what, exactly, the target proportion should be. Stakeholders would probably easily agree to move to 90 percent immature in the catch. Agreement on 70 percent might require some persuasion. If 50 percent is suggested, they may begin to require some serious justification before agreeing. Moving toward a lower proportion of immature fish in the catch will generally require a change in gear (e.g. mesh size), location, or time of fishing, to avoid immature fish. It will generally not be practical to make a change from 90 percent immature to 50 percent immature in one step. Because the cost in reduced catch will be too high for the fisher to bear all at once, reduction will have to be done in steps. In the above example, the first step might be to 85 percent immature, a target probably easily agreed to. Thereafter, each step can be pursued in the context of the benefits derived from the previous step.

Clearly, in this example, the crucial elements of the process will be the ability to monitor the proportion of immature fish in the catch, and the tools and processes required to reach agreement among stakeholders. The former is a relatively straightforward sampling process that should be within the capacity of most small-scale fishery managers. The latter is the area in which most effort will be required in the new fishery management.

6.2.4 Ecosystem-based reference variables

The emerging emphasis on integrating ecosystem, environmental, or ecological objectives into management that was introduced in section 1.2 of Chapter 2 and also considered in section 3.5 of Chapter 3, means that we need to find the appropriate reference variables for these types of objectives. It also leads to the question of whether a large part of fishery assessment should be an environmental impact assessment that would identify the relevant reference variables and indices. As Chapter 3 showed, US fishery management plans must contain an environmental impact statement.

Strictly, ecosystem reference variables should be based on "ecosystem emergent properties," properties of the whole ecosystem but not of its component parts. Examples of emergent properties are biodiversity and resilience. In practice, these properties will not be easy to use as reference variables. What most practitioners refer to when they speak of ecosystem management is management measures aimed at preserving various ecological relationships that are believed to be important for sustained ecosystem functioning. This is the thrust of the ecosystem approach developed by the US Ecosystem Principles Advisory Panel (EPAP 1999). In fact, fishery managers have been taking ecological relationships into account from the earliest attempts at management.

Viewed from the simpler perspective of attempting to manage ecological relationships, it becomes easier to identify variables that relate to linkages between the fishery being managed and its environment. Thus, indices that relate to abundance of predators, prey, competitors, critical habitats, disease outbreaks, and blooms of toxic plankton are appropriate. The ecological perspectives on exploited fishery systems provided by Caddy and Sharp (1986) are an excellent source of ideas for reference variables that relate to ecosystem functioning. Many of us consider the incorporation of some of these into management to be standard practice under the heading of multispecies management; for example, taking into account the needs of predators when harvesting prey species.

A new direction for fisheries management is the consideration of fishing's effects on non-target species, either as bycatch (Alverson *et al.* 1994) or through the destruction of habitats (Dayton *et al.* 1995; Jennings and Kaiser 1998). Because they are impacts, there will inevitably be the need to view their management from the perspective of setting limits. Here, as in the assessment of the fishery resource itself, funds and expertise for the research needed to technically determine acceptable limits are often unavailable to managers of small-scale fisheries. Thus, similar principles to those proposed for the resources will apply: find reasonable, readily observable indicators for the ecological characteristics that are of concern and try to reach agreement among stakeholders regarding what the limits on these should be. The realization that the ecosystems supporting fisheries have been significantly changed by fishing and other human activities, and that the changes are continuing to take place, has led to the statement that rebuilding ecosystems should be a main goal of fisheries management (Pitcher and Pauly 1998).

6.2.5 Consensus on reference points and directions

In the preceding section, we argued that a variety of indicators of fishery performance may be useful to the fishery manager in setting targets, limits, or directions for fishery management and in monitoring progress toward them. Some of these may be related to existing models that can be used to estimate optimal points. Many, however, do not have a basis in quantitative models or are only loosely related to them. In these cases, it may not be feasible to provide quantitative assessment of the optima. Nonetheless, they may be well-founded in common knowledge or qualitative conceptual models. In these cases, for the targets or directions to be adopted, there needs to be an emphasis on reaching agreement among stakeholders on what needs to be done, even if the solution is imperfect; the stakeholders will also need to agree to start to carry out that solution.

The participatory management of a reservoir fishery in northeastern Brazil is an example of how fishers discussed problems, agreed on solutions, and formulated a plan for their fishery (Christensen *et al.* 1995). The fishers identified 13 problem areas: capture of fish during the spawning migration, use of fine-mesh gillnets, unclear land distribution rights, and the need for better fisher participation. The fishers formulated a plan with three measures that they believe will improve the fishery: prohibition of fishing for 15 days after the beginning of the spawning run, allocation of three bays as protected areas, and a ban on small-mesh nets during a part of the year. Scientists with a background in stock assessment would query the technical basis for these measures. Is the mesh size large enough? Is the protected area a sufficient proportion of the entire area? These are valid questions, but answering them may require years of research. Action taken on the basis of the fishers' knowledge and concerns does not preclude research if funds are available. However, there is also the potential to learn from the actions, and to adapt the measures or add new ones in response to what is learned.

The Jamaica conch fishery is another example in which a management plan based on limited preliminary information was implemented by virtue of agreement among stakeholders and resulted in curtailment of the rapid growth in landings for a new fishery (**Box 6.1**).

BOX 6.1 A PRECAUTIONARY, COMMON-SENSE APPROACH TO THE PEDRO BANK CONCH FISHERY, JAMAICA.

Throughout the Caribbean, queen conch fisheries have a history of severe overexploitation. In the late 1980s, the Jamaica fishing industry discovered large numbers of conch on the central plain of Pedro Bank, 150 km south of Jamaica, at depths greater than could be accessed by free-diving. In collaboration with Jamaican entrepreneurs, surplus conch fishing effort in the form of large commercial diving vessels and experienced divers from other parts of the Caribbean quickly began to exploit this resource. The queen conch resource on Pedro Bank appeared to be threatened with the same fate as conch resources elsewhere in the Caribbean.

Rapid, common-sense precautionary action was able to bring exploitation of Pedro Bank conch under control (Aiken et al. 1999). The sequence of events is shown in the figure.

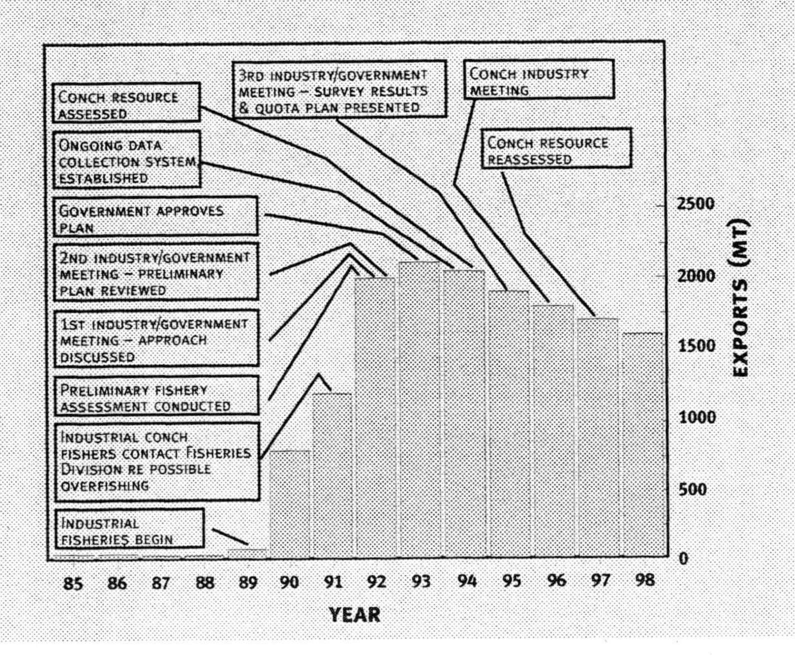

When conch fishing first began in the deep areas of Pedro Bank in 1989, conch exports increased rapidly to just over 2 000 tonnes in 1993. In 1991, members of the fishing industry expressed concern about the fate of the conch resource. The Fisheries Division sought help from the regional fisheries organization (CARICOM), which carried out a rapid assessment of the fishery that included: small-scale and commercial capacity, a review of yields from other Caribbean conch fisheries, and existing information on depths and habitats on Pedro Bank. The assessment indicated that a total annual yield of 600 to 800 tonnes could be expected. Scientists, managers, and industry representatives met to discuss the information. Industry agreed that the annual catches being taken at the time were too high, and that management was urgently needed to avoid a collapse of this fishery. From options discussed at the meeting, scientists and managers put together a precautionary plan that was discussed at a second meeting later that year. The plan was accepted and the major participants agreed to abide by its measures, which included quotas and a closed season, until a more accurate assessment of the stock could be carried out. This rapid response to the growth in the fishery was able to halt the dramatic increase in landings and bring about a slow decline toward a sustainable level.

The first survey-based estimate of MSY for Pedro Bank conch in 1994 using industry funds found that conch density was considerably higher than on other Caribbean grounds and the stock consisted of a large percentage of conch more than five years old. Because the estimated MSY of 1 818 tonnes/year, though almost double the preliminary estimate, was still lower than the current catches, the planned reduction in catches continued. In 1997, a second industry-funded assessment provided a lower MSY estimate of 1 350 tonnes/year, so the plan to reduce catches was kept in place.

In summary, the management process outlined in **Figure 6.4** emphasizes agreement in implementing fishery specific management plans for small fisheries. The process allows implementation to proceed to a conclusion, even in data-limited situations, using the best available information and common sense. As previously noted, this is consistent with the various instruments promoting responsible fishery management and the precautionary approach.

The techniques and methods that can be used with groups to reach consensus in setting objectives were described in Chapter 3 (**Boxes 3.1** and **3.2**). These, and similar methods, will be useful in getting agreement on all aspects of fishery management shown in **Figure 6.4**. The small-scale fishery manager will need to ensure that these skills are available, whether in the government, among the fisherfolk stakeholders, or from an objective third party.

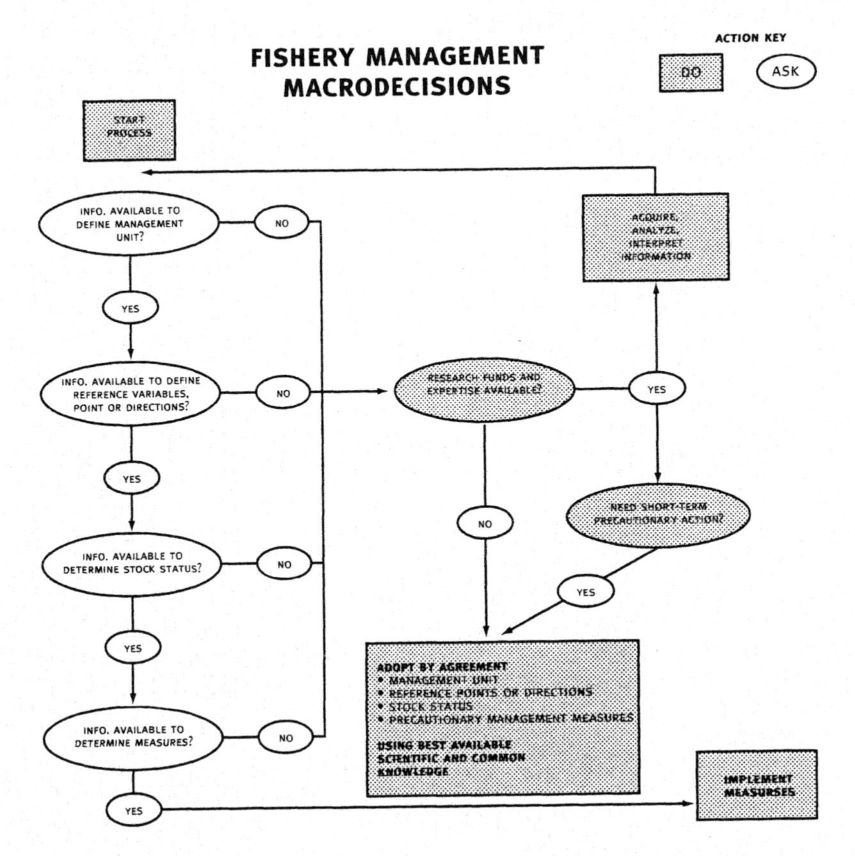

Figure 6.4 A decision-making sequence for fishery management that emphasizes the need for agreement, even when the fullest desirable information may not be available.
Lack of information may be a short-term condition, exisiting only while data collection and analysis is in progress, or a long-term-condition due to lack of funds and expertise. In either case, management must proceed using the best-available knowledge.
Source: adapted from Caddy and Mahon (1995).

6.2.6 Sets of decision rules and procedures

Approaching management in terms of targets, limits, and directions will lead to sets of rules such as in the example in section 6.2.2. above. If there are multiple targets, limits, and directions, these rules may become quite complex and require formal structure. Whatever the case, the rules must be clearly specified so stakeholders can know in advance what criteria will be used to make decisions. The process may even go as far as to specify the kinds of data and analyses that will be used in arriving at the decision. Cochrane *et al.* (1998) refer to this entire package as a "management procedure," noting that it was developed for the International Whaling Commission, although due to the moratorium on whaling it was never applied. These authors provide an example from the pelagic fisheries of South Africa, where the fisheries for two species — sardine and anchovy — are linked. In this case, industry participated in developing the procedure, which considers total catches, interannual variability in catch, and the risk of stock collapse. The procedure specifies the basis for catch quotas, and for mid-season adjustments to these quotas. Whereas this procedure is based on rather sophisticated analyses, the concept can also be applied to situations where data and expertise are limited and the criteria are derived by agreement. Caddy (1998) describes new limit reference points can be combined to provide sets of rules that determine management action. Transparent decision-making contributes to the building of trust.

Now that we realize that fisheries management will require information of several types from a variety of sources — scientific, traditional, local, and administrative (see also Chapter 4) — the challenge is to find ways of combining this information for a particular fishery into a transparent, understandable, communicable form. As McConney (1998) points out, can be achieved by bringing stakeholders together and, with the aid of group process techniques, developing a "common science" approach that incorporates both fishers' knowledge and science.

For more complex situations, and to increase the objectivity in decision-making, there may be the need to employ expert systems. The application of these to fisheries is only now being developed and tried (**Figure 6.5**). Mackinson and Nottestad (1998) point out: "Typically, expert systems are used to solve problems that cannot be solved using a purely algorithmic approach: those that have irregular structure, contain incomplete, qualitative or uncertain knowledge, are considerably complex, and where solutions must be obtained by reasoning and available evidence and sometimes making best guesses." This description clearly applies to fisheries systems. Computer-based "expert systems" that are capable of using a variety of quantitative and qualitative information to deal with problems of the type just described are presently the subject of research. When they are more widely available, and user-friendly, they should be useful to fisheries managers. They will address some of the problems of small-scale fisheries that we have identified; namely, the need to use a wide variety of information of various types and quality in a formal structure or procedure. However, the cost of building them, and the expertise required to operate them, may place them in the same category as conventional stock assessment: unaffordable, or their cost may be unjustifiable for small-scale fisheries. While this interesting development unfolds, one still has the alternative that we refer to throughout this book: the establishment of a framework within which stakeholders can discuss options and reach consensus on the measures to be taken.

Attempts to implement the framework that we refer to above may reveal a gap in fisheries management, the filling of which will require a new type of skill or capability. The role to be played by the person that will fill this gap is that of mediator/synthesizer, an objective and knowledgeable third party who can cope with inputs from both scientific/technical stakeholders and industry stakeholders.

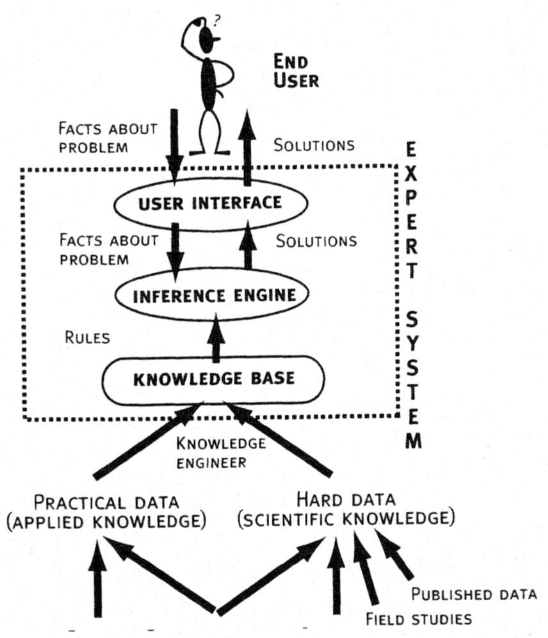

Figure 6.5 The components of an expert system model.
Knowledge base: Typically, knowledge is stored in the form of rules (if a certain situation occurs, then a known outcome is likely), although more complex data structures can be used. This information is gathered from experts and put into the system by the "knowledge engineer."
Inference engine: This compares rules against known facts given by the user to determine if new facts can be inferred; that is, given the conditions, it searches for possible solutions.
User interface: This links the system and the user. It is designed to provide explanations of the actions of the system so that the user can get help or ask, "Why?"
Source: Mackinson 2000

This is the interface between science, industry, and management that has frequently been talked about but is elusive in fisheries management.

6.2.7 Considering Risk in Management Decisions

Managers often consider the various types of risk that may be associated with management choices. As mentioned in Chapter 2, the trend in conventional assessment is toward quantification of uncertainty and risk. Even though the manager of a small-scale fishery may not have the resources to quantify uncertainty, awareness of the various sources of uncertainty in fisheries management systems can be useful in choosing management measures. **Box 6.2** describes the types of errors and associated uncertainty found in fisheries management.

Box 6.2 Five types of uncertainty that arise from an imprecise knowledge of the state of nature.

Measurement error in the observed quantities such as the catch or biological parameters. For example, sample surveys give rise to standard statistical problems of sample size and representativeness; difficulty in accounting for discarding continues to bias landing statistics in many fisheries. In log book and reporting systems there is often misreporting; in quantifying effort, there are often hidden increases in the fishing power of boats. These problems have been an issue for fishery statisticians and assessment scientists for several decades.

Process error due to the underlying stochasticity in the population dynamics, such as the variability in recruitment. The natural variability associated with fish production systems can be enormous. Environmental variability, the largest source of process errors, usually manifests itself as recruitment variability. In short-lived populations, this can result in dramatic fluctuations in adult biomass. Little success has been achieved in the prediction of environmental conditions, or the responses of fish populations, sufficiently far in the future to be useful to management. Since fish stocks become more susceptible to environmental variability as exploitation increases, management can have a direct effect on uncertainty, so reduction of uncertainty may be chosen as a management objective.

Model error due to the mis-specification of model structure. This is seldom evaluated because the data required to distinguish among different models are not available. Studies on the relative performance of various model formulations, such as the Schaeffer and Fox production models, suggest that they may provide substantially different answers using the same data. Evaluating model error requires large amounts of data and considerable expertise.

Estimation error resulting from any, or a combination of, the above uncertainties. This is the inaccuracy and imprecision in estimates of abundance or fishing mortality rate. Owing to the sequential nature of assessment, estimation errors occur at several stages and are propagated through the process. Attempts to quantify estimation error use the variability in measured parameters. However, procedures often use assumed or unmeasured inputs for which there is no information on variability.

Implementation error resulting from variability in the implementation of a management policy or advice. This, usually outside the scientific component of fisheries management, is very much in evidence but has been little studied. Implementation error is largely the failure to control exploitation by whatever measures have been adopted. The reasons are many and interrelated; for example, ineffective surveillance and enforcement, lack of judiciary concern when hearing cases, and participants' failure to support measures because they lack opportunity for input or because they disagree with the measures. In management systems based primarily on advice from biological assessments, failure to incorporate, or incorrect incorporation of, non-biological information also contributes to implementation error. Managers and their technical advisors may know about these problems, but it may be impossible for them to quantify the uncertainty, except in retrospect.

See Rosenberg and Restrepo (1994) and Caddy and Mahon (1995) for more detail on uncertainty.

Errors include not only statistical error in detecting stock status and environmental trends or errors in population analysis but also wrong decisions and the ineffectiveness of a management framework. These are grouped together under the heading "implementation error." Implementation error, whose effects may outweigh all others combined, is not amenable to scientific analysis. Rather, it lies in the domain of human organization and systems management.

To incorporate risk into management decision-making, one must go beyond the probability of occurrence of particular events and consider the degree to which the events are undesirable; that is, the cost or impact of the event.

It is useful to think in terms of two categories of risk (Mace 1994):

- The risk of not achieving a TRP; and
- The risk of exceeding an LRP.

The risks of not achieving a TRP are usually defined in terms of the short-term reduction or interruption of the flow of benefits to fishery participants and consumers. The risks of exceeding an LRP range from stock decline to collapse, damage to associated species, ecosystem destabilization, and long-term loss of earnings, including intergenerational effects.

There are no standard methods for communicating uncertainty and risk to fishery decision-makers (Rosenberg and Restrepo 1994). Basic statistics provide a variety of means of communicating variability, which can be used to indicate the uncertainty associated with a particular estimate or the probability of occurrence of an undesirable event. The method chosen to communicate uncertainty and risk to managers depends on technical capability. In most developing countries, it will be important to relate the uncertainty to well-known characteristics of the fishery, such as amount of catch, rather than to a fishing mortality level estimated using a complex analytical process. For example, a simple graphical presentation was used in the eastern Caribbean to communicate trends in yield of flyingfish, catch rates, their variability, and the probability of undesirable events being brought about by increasing fishing effort (**Figure 6.6**).

The current focus on the quantification of uncertainty and risk in natural resource management requires a considerable amount of information and expertise. Fishery advisors and managers must note that subjective views of risk, based on the experience of participants, can also be applied in management. Most informed fishers and managers would agree that there is an unacceptably high risk that

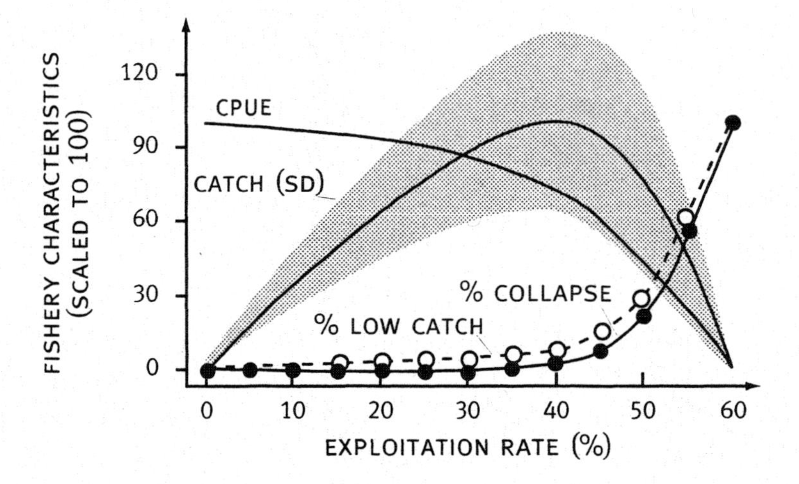

Figure 6.6 A summary of fishery characteristics based on a stock-recruitment simulation for eastern Caribbean flyingfish, indicating the risk of undesirable occurrences associated with increasing levels of exploitation.
These undesirable occurrences, which have been defined subjectively, include: variability in catch, and, by inference, catch rate; probability of years with "critically low catch," defined as annual catch < 30 percent of current average catch; and "collapse," defined as critically low catch for four or more years.
Source: Mahon 1986

uncontrolled fishing on grouper spawning aggregations will lead to extinction of the aggregation, and that access to aggregations should be controlled. No assessment of the particular stock is required for management action in such a clear case. The data to estimate an optimal escapement may not be available, nor, due to discounting, may it be perceived as economically feasible to acquire and analyze, but sustainability can be achieved by limiting access as a precautionary measure.

6.2.8 Adaptive management

Does the adaptive management introduced in Chapter 2 have practical applications for small-scale fisheries management? In the Pacific island country of Vanuatu, the Fishery Department faced a conservation problem and initiated a *Trochus* management program (Johannes 1998b). The villages adopted the department's advice about harvesting frequencies and the need for a fishing-free period between harvests. They were rewarded with improved yields and incomes. News about their success soon reached other villages and, within a few years, many more villages were managing their *Trochus* stocks. As well, some villages started to adapt *Trochus* controls to other species, including some finfish, lobsters, and octopus. All but one of 27 villages implemented their own community-based marine resource management measures. Also notable was the diversity in practice: no two villages had exactly the same set of conservation measures. Each set was tailored to the specific socio-cultural and biophysical circumstances.

Thus, the department's modest efforts in a few villages spread to a much larger area through a learning-and-demonstration effect. All of this was accomplished without biological data other than *Trochus* growth rate data for the region and the villagers' own information in the form of increased income from *Trochus* management. Johannes (1998b) considers the experiment to be an indigenous version of adaptive management, involving trial-and-error learning by experimentation followed by dissemination of local knowledge. Although it is not scientific adaptive management, such trial-and-error learning no doubt played an important role in the folk management of marine resources centuries before scientific managers appeared on the scene.

Adaptive management relies on deliberate experimentation followed by systematic monitoring of the results, from which managers and resource users can learn. Many small-scale fisheries are in or are facing a crisis of declining catch and incomes. Livelihoods are threatened. It is not possible, with available resources and time, to conduct stock assessments and collect the other information needed to develop a management plan and strategy. Something needs to be done now.

Through adaptive management, the available information is collected, there is consultation among fishers and managers, an action strategy is agreed upon, and some action is taken. This action is monitored, information is analyzed, and lessons are learned. The action, depending on its level of success, is adjusted as necessary. When fisheries are at risk, it is better to try something and learn from it than to do nothing at all: that is the approach of adaptive management. Thus, the identification of management reference directions (section 6.2.2) can be seen as the starting point for an adaptive approach through which the desired target or end-point can be determined. These policy/experiments can be initiatives like Vanuatu's *Trochus* conservation. Certainly, recognizing coastal community jurisdiction over fishing territories is a major policy/experiment that has significance for many areas. Similarly, a great deal of background information exists on the Japanese coastal management system, the evolution of local rule-making, and the development of fishing territories and reciprocal access rules (Ruddle and Akimichi 1984; Ruddle 1989). The "dataless" but information-rich management that taps into local knowledge and comparative management information is another learning experience (**Box 6.3**). The adaptive management literature has practical lessons for small-scale fisheries management in two additional areas, one related to learning from mistakes and the other related to maintaining institutional memory.

Box 6.3 "Dataless management."

Johannes has proposed a system of fishery management in which a combination of MPAs and fishers' knowledge replace much of conventional fisheries management research. Because it is not based on biological data of the kind managers use, he calls the approach "dataless management." He starts by pointing out that managing most marine fish to achieve optimum yields is an unattainable dream. Even in high-income countries with extensive management infrastructure, research seldom provides sufficient knowledge to manage for maximum or optimum yields, whether biological or economic.

Tropical nearshore marine fisheries are a clear example of the failure of classical fisheries management. No other fishery involves so many species, habitat types, and users. There is little consensus among fisheries biologists concerning even the basics of the dynamics and management of these fisheries. Yet, Johannes points out, there seems to be a common assumption among many marine biologists and managers that the availability of quantitative information about a natural resource is essential for any kind of management. If this assumption were true, it would be impossible to carry out even a rudimentary protection of these resources simply because the capability to collect and process management data for the vast majority of the resources and the area does not exist. Countries such as Solomon Islands, Papua New Guinea, Indonesia, and the Philippines could not afford scientific research at such levels of detail, but even if they could, it would be grossly cost-ineffective. Thus, alternative approaches are needed.

The alternative approach proposed by Johannes combines MPAs and fishers' local knowledge. MPAs are seen as a tool for precautionary management; that is, not to control resource production but simply to protect and maintain the viability of the resource. Based on research in the Caribbean and the Indo-Pacific, the larger reef fishes of many species tend to spawn in the same locations and seasons. In Palau, for example, more than 40 species spawned in three aggregation sites. Hence, protecting major spawning aggregation sites can help protect many stocks of several species. To include these sites in a MPA, conventional fisheries data are not essential. All that is needed is information on the timing and location of spawning aggregations (which are often well known to fishers) and a local perception that these aggregations are threatened. The protection can take the form of species prohibitions (for the duration of the spawning season) or area closures on the spawning grounds.

Source: Johannes (1998a)

6.3 Management measures

6.3.1 Traditional measures

Documented cases have accumulated, especially since the 1980s, on long-standing community-based management systems. It is becoming clear that these time-tested systems were often based on sound ecological knowledge and understanding, particularly in the Asia-Pacific region, which is rich in traditional knowledge and management systems. Many of these systems have been documented in detail, especially those in Japan and parts of Oceania (Melanesia, Micronesia, and Polynesia) (Ruddle and Akimichi 1984; Freeman *et al.* 1991).

The most widespread single marine conservation measure employed in Oceania was a combination of reef and lagoon tenure and taboos. The basic idea behind reef and lagoon tenure is self-interest and sustainability. The right to harvest the resources of a particular area was controlled by a social group, such as a family or clan (or a chief acting on behalf of the group), who thus regulated the exploitation of their own marine resources. As Johannes (1978) explained, "it was in the best interest of those who controlled a given area to harvest in moderation. By doing so they could maintain high sustained yields, all the benefits of which would accrue directly to them." A wide range of traditional regulations and restrictions applied to resource use. Some of these rules could be attributed to religious beliefs (Johannes 1978) and some to power relationships and regional differences in systems of political authority (Chapman 1987). But by and large, reef and lagoon tenure rules served both conflict resolution and conservation, directly or indirectly, and operated as institutions for the management of common property resources (see Chapter 7 for more detail).

These management systems provide insights about sustainability in general. They also provide biological knowledge that can be used in scientific management systems. A telling example of the level of detail available from indigenous knowledge is provided by Johannes, an expert on tropical reef-fish ecology. When Johannes was working with fishers in the tiny Pacific archipelago of Palau in the mid-1970s, he obtained, from local fishers, the times and the precise locations of spawning aggregations of some 55 species of fish that took the moon as their cue for spawning. This local knowledge amounted to more than twice as many species of fish exhibiting lunar spawning periodicity as had been described by scientists in the *entire world* at that time (Johannes 1981).

One of the first international projects on traditional ecological knowledge concentrated on coastal management systems from around the world, pointing out the variations and similarities in the methods devised by peoples of very different areas and cultures (Johannes *et al.* 1983). Their examples included:

- The *valli* (or *vallicoltura*) of the Venice region, Adriatic Sea;
- The *cherfia* of North Africa, installed at the mouth of lagoons, similar to the *lavoriero* of the Italian coast, with variations in Portugal, Greece, and Turkey;

- The *acadja* of West Africa, which involve immersing piles of branches in the shallow parts of the lagoon, as also done in brushpile fisheries of Bangladesh and Sri Lanka; and
- The Indonesian *tambak*, which were originally brackish-water fishponds, usually installed in delta systems and associated lagoons.

The above examples are only a small sample of such systems; additional types have since been documented. Some of these coastal systems, of particular interest to managers, are excellent examples of the application of a pre-scientific ecosystem view (Berkes *et al.* 1998). One example is from Indonesia, where traditional systems combined rice and fish culture, with nutrient-rich wastes from this rice field fishery system flowing downstream into brackish water aquaculture systems (*tambak*) and on into the coastal area, enriching the coastal fishery (Costa-Pierce 1988).

6.3.2 Conventional and new fisheries management

Much has been written on tools or control measures for fishery management. Some were developed out of common knowledge and long-term observation in traditional management systems (**Table 6.2**). These and others developed in the last 100 years as part of conventional fisheries management, such as those described by Beddington and Rettig (1983) and Hilborn and Walters (1992), comprise the present tool box for fisheries managers (**Table 6.3**).

Table 6.2 Frequency of occurrence of traditional fishing regulations in 32 societies from around the world.

Type of regulation	Frequency
Areas (community controlled)	30
Limited access	16
Technology	12
Seasonal limits	10
Protect breeding stock	9
Protect young	8
Conservation ethic (of individuals)	6
Size limits	4
Overcrowding	3
Quotas (on catch)	1
Other	8

Source: Wilson et al. 1994

TABLE 6.3 AN OVERVIEW OF THE MAIN METHODS USED FOR CONTROL OF FISHERIES (I/O = INPUT/OUTPUT CONTROL, D/I = DIRECT/INDIRECT CONTROL).

METHOD	AIM/EFFECT	COMMENTS
Licensing, limited entry (ID)	Licensing is the only way to directly limit the number of participants in the fishery.	Licenses can be used as a means for recovering some revenue from the fisheries. Licensing alone is seldom enough to control the amount of fishing effort.
Effort limits (ID)	These direct limits to the number of units of effort; for example, hours fished, traps pulled, or trawl sets.	Limiting effort in this way is more direct, but fishers usually find ways of getting around effort limits by increasing aspects of effort that are not limited; i.e., larger traps or larger boats.
Closed season (II) Closed area (II)	These aim to protect a specific part of the stock known to occur in or at a particular place or time; usually spawning or young fish. May also be used to control total effort by eliminating fishing from a particular area of the stock or period of the year.	When used as a means of controlling total effort, fishing usually increases in the open area and at the open time of the year. Thus, reduction of effort is not directly proportional to the closed season or area. Closed seasons are easier to monitor than closed areas, unless the latter are very large.
Gear restrictions (II)	These usually aim to control the size or species of fish caught; for example, by regulating the mesh size used in nets or traps.	Although the relationship between gear and size of fish caught is imprecise, gear restrictions can be monitored by inspection ashore.
Catch quotas, total allowable catch (TAC) (OD)	Quotas directly limit the amount of fish taken from the stock to that corresponding to the target reference point. TAC is the simplest form of catch quota.	Catch quotas vary with the abundance of the resource and must thus be re-estimated at regular intervals. This requires substantial amounts of detailed data. Regulation by catch quotas also requires that fish landings be monitored on a real-time basis so that the fishery can be closed by the catch when the quota has been taken. A single TAC often results in a race for the quota and, consequently, overcapitalization.

(continued)

Table 6.3 concluded

Method	Aim/Effect	Comments
Industry quotas (ID)	The TAC is divided up among participants in the fishery.	Individual fishing companies can manage the way in which they take their share in order to optimize their economic return. The equitable distribution of quotas among participants is usually difficult and contentious.
Individual transferable quotas (ITQs) (OD)	This is a form of industry quota in which the quotas may be transferred, sold, or traded.	ITQs facilitate the operation of normal market effects in the fishing industry. More efficient companies can buy quotas and so increase their share of the resource. A basic proportion of ITQs are given out on a long-term basis so that companies may plan their operations. Remaining quotas are distributed or sold each year, with the amount becoming available being dependent on the abundance of the resource. May lead to monopolies.
Size limits (OD)	This directly limits the size of fish landed in order to reduce growth of over-fishing and to ensure that immature individuals are not caught.	Shore-based monitoring of size limits will often lead to discarding of smaller sizes at sea. Because discarded individuals usually die, this defeats the purpose of the regulation.
Taxes or tariffs (OI)	Taxation on the fish landed is one means of reducing the amount of fish caught.	This increases the cost of fishing, thus shifting the cost and revenue equilibrium toward lower effort

In the new approach that we are describing, most of the tools that managers could use to move a fishery in the desired direction, or away from a limit point, are in the existing toolbox, perhaps to be applied in a new way. The main difference between the old and new approaches is the way of reaching decisions on when and how to apply the tools.

6.3.3 Rights-based fisheries

As will be further discussed in Chapter 7, Hardin's (1968) model of the "tragedy of the commons" has been used to help explain the overexploitation of fishery stocks around the world. Economists have pointed out that governments often use the wrong approach to deal with problems that arise from a "market failure." Command-and-control regulations used in an attempt to correct the problem through legal and administrative means do not dissolve the market failure.

This has led to a market-based perspective that suggests that governments should change the incentive structure in order to bring private and public interests closer in line, restoring the workings of a "perfect" market. Advocates of a market-based perspective see the lack of individual property rights as a market imperfection in the fisheries.

This perspective led to the development of property rights regimes to regulate access and effort. In temperate developed countries, biological fisheries management used biological knowledge of fish stocks to set Total Allowable Catches (TAC) designed to restrict exploitation of the stock at or below the TAC (Beddington and Rettig 1983). Economic fisheries management included economic instruments to regulate exploitation at the TAC, thereby extracting maximum economic rent. The introduction of Individual Quotas (IQs) and Individual Transferable Quotas (ITQs) were steps toward individual property rights that, ideally, help to restore the workings of the market mechanism, address problems of governance, and add objectivity in adjusting fishing effort.

These management models, however, seem better suited to temperate regions with discrete single-species fisheries, and therefore calculable TAC, than to the multispecies, multi-gear fisheries of many tropical countries. These models have limited applicability to tropical fisheries because of the large amount of information that managers need to implement them, the wide variety of fishing gears used in the tropics, and managers' limited ability to control access of fishers, both full- and part-time, to the tropical fishery.

Quotas, particularly individual transferable ones, have been promoted as appropriate rights-based management tools for several fisheries in developed countries. They are also being introduced to some developing countries, such as in the Jamaican conch fishery (**Box 6.1**), but there are several reasons why quotas are problematic, especially for small-scale fisheries (**Table 6.4**). However, other tools based on fishing rights have been used for centuries in the traditional fisheries of many communities, particularly in the small islands of the Pacific.

Table 6.4 Reasons why quotas are problematic for small-scale fisheries.

Quota and fishery features	Issues that may confront many small-scale fisheries
Quota busting	Poor enforcement resulting in quotas often being exceeded
Data fouling	Inaccurate catch reporting due to cheating or complexity
High variation stocks	Widely variable year classes, abundance, availability, etc.
Short-lived species	No clear relation between stock and next year's recruitment
Flash fisheries	Season too short to be monitored for management
Real-time management	Precise control of effort difficult with dispersed fisheries
High-grading	Market strategy of discarding low-value fish encouraged
Multispecies fisheries	Not possible to set optimal catch or effort for a complex of species
In-season variation	Declining abundance in-season resulting in a race at the start
Information for TAC setting	Information base inadequate for setting the TAC with precision
Transitional gains trap	Unpopularity of taxing the gains of initial beneficiaries
Industry acceptance	Low acceptance if initial allocations are seen as inequitable
Spatial distribution of effort	Overexploitation of high-yielding grounds due to patchiness
Quota concentration	A few companies or rich people buying out many small fishers
Social and economic change	Affecting society more than many other management tools

Source: Copes 1986

Several of the constraints and challenges facing small-scale fisheries managers have been identified before, and the table below relates these to quota systems. Several of them are also problems for large-scale and developed-country fisheries, but in these cases the nature of either the fisheries resources or management capacity makes them less critical.

Other rights-based management tools have focused on quotas for groups such as communities or fishing-industry organizations rather than individuals or companies. The most durable systems have concerned access to the fishing area or gear rather than to the amount of catch. The common feature is the decentralization of control over exploitation, usually by devolving power that was centred upon the state. In the past, this control was often community-centred and at the origin of the rights-based system integrated with the social and cultural practices of the resource users. Now, formal recognition by the state is also necessary for full legitimization and acceptance by the wider society and by outsiders who may seek to impose different values.[2]

It is important for small-scale fisheries managers to investigate the existence or introduction of rights-based fisheries management approaches. There is ample evidence that such approaches can be successful and sustainable when appropriate to the human and ecological systems.

6.3.4 Obvious measures

When a fishery assessment has been carried out, the conclusion may sometimes be drawn that the most serious problems faced by the fishery, or even the resource, cannot be addressed through stock assessment. Following are two examples pertaining to coral-reef fishing. The first is about destructive practices or gear such as dynamite, chemicals, and trawls (McManus 1997). In these cases, the damage to resources and their habitats is such that any standard assessment approach would be inapplicable, or at least superfluous. The problem is obvious, and the solutions are institutional, not technical. The second example is the targeting of spawning aggregations. These are common among several families of tropical fishes, particularly groupers and snappers (Domeier and Colin 1997). Here, the fishery has the potential to exert mortality in ways that cannot be easily measured using standard assessment approaches or controlled using standard measures. Here again, the problem is obvious and diagnosis requires minimal technical expertise or science. The solutions are institutional, comprising rights and rules that control behaviour, and are often as difficult to implement as other measures, owing to the resulting effects on fishers.

6.3.5 Ecosystem-based measures

The impact of environmental degradation from both fishery and nonfishery activities on the ecosystems that support fisheries, particularly inland, coastal, and inshore fisheries, is increasingly recognized as the major fisheries management problem (Dayton *et al.*1995). Separating these impacts on exploited resources from the direct

[2] See also papers presented at FishRights99 Conference, 14–17 November 1999, Fremantle, Australia.

effects of fishing mortality may be one of the major challenges of fisheries management planning. Since most small-scale fisheries are near shore, nonfishery human impact is usually a more important issue in their management than in large-scale fisheries. Consequently, different types of management measures are likely to be useful, depending on distance from shore (Caddy 1999). For inshore and inland fisheries, habitat conservation, rehabilitation, and enhancement are commonly used management measures.

Although this is an emerging field, it appears that ecosystem-based measures will be variations of standard measures based on ecosystem criteria. For example, areas may be closed to protect habitats; quotas of prey species may be set to ensure adequate forage for predators; and predator quotas may be set to ensure that predator depletion does not lead to explosions of prey populations that are released from predation pressure. For discussion of and guidelines on managing fisheries that involve predator and prey species, see Christensen (1996).

6.3.5.1 Marine protected areas

As discussed in Chapter 2, marine protected areas (MPAs) have the potential to play a significant role in aquatic resource conservation. They can be used to set aside representative areas for the conservation of biodiversity. They can also reduce conflicts between fishers and other users by providing areas where non-fishery users can pursue nonconsumptive uses of the resources. On the other hand, the extent to which they can enhance fisheries is less clear. It has been noted that MPAs should be considered a necessary but not sufficient component of a small-scale fishery management plan. One of the main concerns about relying too much on MPAs is that they simply displace fishing into adjacent areas, leading to extra depletion there (Fogarty 1999). This, in turn, increases the difference in abundance between protected and exploited areas, which may increase emigration of fish from the former to the latter, thus reducing abundance in the protected areas.

Although there are gaps in our scientific knowledge about MPAs and how they function, many of the ideas of species and area protection may be found in long-standing traditional management systems. For example, traditional closed areas in Oceania may have served many of what we could now consider scientific management functions, and the same can be said for bans on critical life-history stages. As well, some traditional systems protect spawning aggregations or spawning runs. In the ancient salmon management systems in the Pacific Northwest of the United States and Canada, protection was not based on a total area closure or a species ban but on the escapement of a critical population of spawners upstream before fishing was allowed (Swezey and Heizer 1977).

In proposals to manage small-scale fisheries in the Pacific with minimal data requirements, a combination of local knowledge and the use of studies on similar fisheries in other locations would replace conventional management data (**Box 6.3**). A crucial point in the plan is to include key spawning areas in MPAs. In practice, though, such measures may not work smoothly because spawning sites may not be located in areas where monitoring and enforcement are feasible. In any case, although protecting the reef-fish resource in the manner proposed by Johannes will not result in "optimum" management, it is preferable to the only real alternative in most cases: *no management at all.*

Johannes (1998a) points out that carrying out such "dataless" management will not be easy. Success will depend on managers being in tune with fishers and committed to working with fishing communities. MPAs that are to protect the viability of major fish populations need to be designed with those objectives in mind, not to carry out the requirements for arbitrarily designated protected areas. They will need the local fishing communities' support and help in monitoring and enforcement. Most importantly, the viability of such MPAs will depend on government recognition and support from local marine tenure systems that regulate the use of the commons.

The worldwide experience so far with MPAs is that many of them are "paper parks," with insufficient funding, infrastructure, and controls. One solution is to enlist the help of the local community to enforce conservation, as documented in St. Lucia for the edible sea urchin resource (Smith and Berkes 1991). The National Research Council report comes to a similar conclusion: "In all cases the involvement and support of local fishers were a prerequisite for any success of the reserves . . . . Enforcement was a problem that could be solved only when local fishers were sufficiently committed to the reserves and sufficiently concerned about threats to their resources that they were willing to act together to enforce the rules and prevent poaching" (NRC 1998, pp. 89–90).

One of the reasons that parks frequently do not perform as intended is that many are set up with the hope that they will serve several purposes: tourism, biodiversity conservation, and fisheries enhancement — purposes that are not always compatible. Consequently, one must pay careful attention to criteria when establishing MPAs. Salm and Clark (2000) provide an extensive review of marine MPAs, giving considerable guidance about the criteria that should be used in establishing them (**Table 6.5**). Appeldoorn (1998) and Pitcher (2000) also provide guidance about criteria and expected fishery and non-fishery benefits for MPAs.

Table 6.5 Criteria for the selection of protected areas.

Criteria	Comments
Social	
Social acceptance	Degree of local acceptance and concern
Public health	Reduce pollution
Recreation	Provide opportunity for enjoyment of area
Culture	Special religious, historic, artistic value
Aesthetics	Land and seascapes of special beauty
Conflicts of interest	Degree of disruption to existing users
Safety	Extent to which area is hazardous to potential users
Accessibility	Ease with which it can be reached
Research/education	Extent to which site is useful for these
Public awareness	Extent to which site will increase this
Conflict and compatibility	Extent to which site may help to resolve existing conflicts
Benchmark	Can site be a control site for monitoring?
Economic	
Importance to species	Can site contribute to sustainability of economically important resources (e.g. reefs)?
Importance to fisheries	Similar to above
Nature of threats	Extent to which use of site will cause activities to change detrimentally on adjacent sites
Economic benefits	Direct benefits to local and national economy
Tourism	Potential value of site for tourism development
Ecological	
Biodiversity	Is site particularly high in diversity, or unique?
Naturalness	Is site natural (important if protecting is priority) or degraded (important if restoring is priority)?
Dependency	Extent to which area is critical habitat for species
Representativeness	Is site one-of-a-kind or representative of wider array of habitats?
Uniqueness	Similar to above
Integrity	Extent to which area is a functional ecological unit
Productivity	Extent to which productivity in the area benefits humans
Vulnerability	Susceptibility to degradation
Regional	
Regional significance	Extent to which area is characteristic of its region
Subregional significance	Extent to which area fills a gap in regional conservation
Pragmatic	
Urgency	Need for immediate action
Size	Area needed for effective protection versus practicality of management
Degree of threat	If far from threat, likelihood of success is greater
Effectiveness	Feasibility of being managed
Availability	Is site available?
Restorability	Is degradation reversible within current situation?

Source: adapted from Salm and Clark 2000

6.3.5.2 Habitat restoration, creation, and enhancement

Small-scale fisheries, usually located in inland water bodies and coastal areas, are highly dependent on habitats (coral reefs, mangroves, seagrass, wetlands) that are susceptible to human-caused pollution and physical destruction. The restoration of these habitats, particularly those that limit the abundance of a resource at some life-history stage, may be the most important step to increasing stock productivity. The restoration of coastal habitats that have been destroyed by development is increasingly taking place in many developed countries. In some cases, lands that have been filled and reclaimed for agriculture and development have been purchased, at very high cost, and returned, insofar as possible, to their original condition. This trend is based on the realization that many of these habitats are important nursery or spawning areas for fishery resource species. The role of coastal wetlands in maintaining the quality of fresh water that is discharged into nearshore habitats, and thus the integrity of these ecosystems, has been another driving force in coastal wetland rehabilitation.

The need for attention to resources' habitat requirements was recently given legislative weight in the USA through the *Sustainable Fisheries Act*, which, as stated in Chapter 3, requires that an assessment of Essential Fish Habitat (EFH) be included in each Fisheries Management Plan, together with recommendations for EFH conservation.

A number of methods exist to create and enhance aquatic systems. Examples of these are artificial reefs, fish attracting devices (FADs) for pelagics, and casitas for lobsters. Many of these have traditionally been used by small-scale fishers around the world (Kapetsky 1985). (A review of these approaches, which are adequately covered in other publications, is beyond the scope of this book.) However, before implementing them, the small-scale fishery manager must consider whether they are likely to contribute to increased production by the resource or simply increase the availability of the resource to exploitation by aggregating individuals. If the latter, their use must be accompanied by the capability to control exploitation.

Artificial reefs (AR) are structures that serve as shelter and habitat, source of food, breeding area, resource management tool, and shoreline protection. The AR may act as an aggregating device to existing dispersed organisms in the area and/or allow secondary biomass production through increased survival and growth of new individuals by providing new or additional habitat space. In addition, ARs have been considered as a barrier to limit trawling in coastal areas where they may be in conflict in small-scale fishers.

Fish aggregating devices (FADs), items placed in the water to attract fish to aggregate (gather near to them), have been used in Southeast Asia for much of the 20th century, if not longer. FADs are deployed in a variety of environments, from calm waters to rough, high-energy environments (Pollnac and Poggie 1997). They can be constructed from a wide range of materials, from simple line and palm fronds to sophisticated devices with radio beacons. For example, bamboo rafts are traditionally used in Japan. The benefits of using FADs include: 1) increased catch, 2) lowered fuel consumption, 3) accessibility to small-scale fishers, 4) shifted effort from overfished areas, 5) improved fishing vessel safety, and 6) definition of territory and/or inhibition of certain types of fishing. Potential problems include: 1) increased probability of stock depletion, 2) changes in eating habits of attracted fish, 3) lack of monitoring and evaluation, 4) restricted access to the resource, 5) increased conflicts, 6) periodic maintenance and replacement required, and 7) cost of long-lived, high-technology devices (if used).

6.3.5.3 Restocking and introductions

Enhancing fish populations by restocking with young individuals has been most successful in small, enclosed water bodies such as ponds and lagoons (Welcomme 1998). The generally high cost of producing the young for stocking means that this approach has been most cost-effective for recreational fisheries that provide economic returns beyond the landed value of the fish. The few instances of successful stocking programs in the marine environment are in very localized inshore habitats (Blaxter 2000). The small-scale fishery manager should carefully weigh the costs and benefits of any stocking program. As well, the variety of risks, such as of genetic dilution of the wild stocks and introduction of disease, should be considered.

When considering introducing new species, the manager should fully explore the extensive literature that describes the many pitfalls and case histories of unexpected consequences. The *Code of Conduct for Responsible Fishing* provides guidelines for introductions (FAO 1996b).

6.3.5.4 Fish trade and ecolabeling

Preferences that consumers communicate through domestic and international fish markets and trade influence which species are considered targets and which are bycatch of lesser value. The fishing gear type, size distribution, and reproductive condition of the catch, and post-harvest processing are other fishery features that may be determined, directly or indirectly, by trade. Overexploited species often become more valuable due to scarcity, causing the market to exacerbate unsustainable practices. An area of increasing importance in recent times, therefore, is the relationship between trade in fish and fishery products and the sustainable management of fisheries (Deere 2000). An example of this is the Marine Stewardship Council's use of consumer pressure to reduce bycatch of dolphin on tuna longlines or turtles in shrimp trawls.

The *Code of Conduct for Responsible Fisheries* addresses post-harvest practices and trade in Article 11 and in its guidelines for responsible fish utilization (FAO 1998), recognizing the important role of the World Trade Organization (WTO) in formalizing world fish trade. The notion of consumer preferences and certification systems assisting to encourage the purchase of fish from fisheries that have been managed with best practices to ensure sustainability is gaining favour in some quarters (Deere 1999). This trend toward ecolabeling, however, is being monitored closely by developing countries who see it as an opportunity for developed countries to impose hardship through initiatives that are more related to fishery economics than ecology. It is important for developing countries to check that the criteria used for ecolabeling take into account the nature of small-scale fisheries and the alternative means to manage them.

6.3.6 People-focused measures

Successful implementation of fisheries management is now seen to include a variety of measures that engage and inform stakeholders (including the public). Addressing the undesirable social and economic implications of attempts to reduce fishing effort is also an emerging direction.

In several parts of this book, we emphasize the need to inform and build the capacity of fishing-industry stakeholders in order to empower them to participate in fishery development and management (see Chapters 4 and 8). This should be borne in mind as a crucial new direction for management.

6.3.6.1 Public education

Education aimed at the non-fishery public can increase their awareness that they are stakeholders with a right to expect that fisheries will be well managed on their behalf, and that industry stakeholders will observe the agreed-upon measures in return for the right, or privilege, to participate in the fishery. Messages absorbed by the general public, such as those about conservation, can be important for structuring social sanctions and attitudes. A knowledgeable public can also play a role in enforcement, either indirectly through the political directorate or directly by exercising its consumer right not to purchase illegally or inappropriately harvested products.

The role of the public as stakeholder should not be underestimated, particularly when household consumers are the primary purchasers of fish. If the public is aware of the issues, regulations and the long-term effects that breaking the regulations may have on the availability of the product, there is reason to believe that many individuals will choose not to purchase illegally caught fish. If properly informed and supported by the authorities, the public can also play a role in reporting violations. These roles may be strengthened through public education and market-oriented initiatives such as the eco-labeling mentioned earlier. Cases involving sea turtles and marine mammals are well known. However, public perception of the fairness of management also takes into account the opportunities (or lack thereof) for involved fishers to pursue alternative livelihoods.

6.3.6.2 Community development

Occupational multiplicity, a prominent feature of small-scale fisheries, has the consequence that fishery management extends into the domain of integrated community development, whether urban or rural. Many of the management measures previously discussed alter patterns of employment in fisheries and supporting occupations. Community development programs that address alternative employment for fishers, and livelihood planning for part-time fisherfolk, can contribute to the management of fishing effort. Livelihood planning is especially important where households depend on fishery-related income. Extensive excursions into these areas are beyond the scope of this book, but they must be part of a new-style fisheries management. The capacity to contribute to such initiatives, or even initiate them, should be a requirement in a fisheries department that deals with small-scale fisheries. This can be achieved through in-house capacity or, more likely, via close links with agencies specifically responsible for community development. Such development can also increase the quality of life in rural coastal communities through delivery of basic services (for example, health and education) and infrastructure development (for example, roads and communication).

While governments are working to attain sustainable development of coastal and marine resources and to improve the socioeconomic conditions of coastal residents, funds and other resources for these purposes are limited. This is not a new situation, but new action must be taken to deal with these issues. With limited government resources, the fishery stakeholders will need to take more responsibility for finding solutions to their problems. Because the resource users must be involved in making management and development decisions, they need to be educated, informed, and empowered to take action. New governance arrangements for fisheries and coastal resources must be examined and put into place. Resource management policies must shift from a resource exploitation orientation to one of more holistic conservation and human resource management. In order for socioeconomic development to be sustainable, attention must be given to policies that address issues of food security and people's well-being and livelihood, not just regulatory fisheries management.

Mixed with policies concerned with resource management and conservation is the need to address problems of poverty, unemployment, and decreasing quality of life in fishing communities. The main brunt of such economic and social distress is borne by women, children, and unskilled fishers, as well as by those unskilled people who depend, directly and indirectly, on the fishing industry. Elements of this prevailing scenario are: high levels of unemployment or underemployment, unavailable alternative or other supplemental employment and livelihood opportunities in the community, a growing population and pressure to find additional fisheries resources, lack of credit and markets, and the paucity of institutional mechanisms to undertake system-wide development.

6.3.6.3 Managing excess fishing capacity

It is now almost universally accepted that many coastal fisheries are overfished. Many small-scale fisheries are home to an excessive level of factor inputs (capital and labour) relative to that needed to catch available fish. Thus, most fisheries can be characterized as having the problem of "excess capacity," "overcapitalization," or simply "too many fishers chasing too few fish." The result is lower productivity of small-scale fisheries, increasing impoverishment of small-scale fishers, and erosion of food security in coastal communities that depend on fish supplies for protein and income.

Because the capital and labour employed in small-scale fisheries are generally use-specific, their exit is often difficult and painfully slow. As long as small-scale fishers can obtain a positive return, they will continue fishing, trying to circumvent any command-and-control regulatory measures such as gear limitations and closure of fishing areas. These measures appear to focus on the resource rather than on the people: the fishers, other resource stakeholders, and the community. Resource managers' action to deal with excess capacity as a major cause of resource overexploitation and environmental degradation reflects a one-sided policy response to the problem. Unless we address the core issue of excessive capacity; that is, by facilitating the exit of labour and capital from the fishery without unacceptably severe social and economic disruption, any regulatory measure or other management strategy will simply be a stopgap measure. People will continue to enter the fishery unless viable alternatives are presented.

As traditional institutions and methods of controlling overexploitation of fisheries fail under the pressures of modernization and market economies, fisheries managers are increasingly aware of the need to develop appropriate policies to facilitate the exit of capital and labour from overexploited fisheries. This growing consciousness of the importance of reducing fishing overcapacity culminated in the FAO Committee on Fisheries' adoption in February 1999 of the *International Plan of Action for the Management of Fishing Capacity*. This instrument calls for states to prepare and implement national plans to effectively manage fishing capacity, with priority to be given to managing capacity in fisheries where overfishing is known to exist. International policy discussions of the fishing fleet overcapacity problem have focused overwhelmingly on industrial fishing fleets, largely ignoring the problems of small-scale fisheries. Developing countries with small-scale fisheries with severe overcapacity are unlikely to prepare effective plans to address that aspect of fishing overcapacity without initiatives to help them analyze the problem and generate new policy options.

The problem of reducing excess capacity in small-scale fisheries in developing countries is much more complex than that of reducing overcapacity in industrial fleets. The complexity in small-scale fisheries is compounded by: growing populations, sluggish economies, fishers' high dependence on the resource for food and livelihood, a paucity of non-fishery employment, increasing numbers of part-time and seasonal fishers, limited transferability of and rigidities in the movement of use-specific capital and labour, and the lack of a coordinated and integrated approach to horizontal economic and community development that blends fishery and non-fishery sectors.

Thus, a reduction of excess capacity implies an increased focus on people-related solutions and on communities. This should involve a broad program of resource management and economic and community development that emphasizes access control and property rights, rural development, and linking of coastal communities to regional and national economic development. This new management direction needs to address coastal communities' challenges, including employment and income, food security, better quality of living, and delivery of community services. We must go beyond the "common" solution, which is to give fishers "pigs and chickens" as a supplemental livelihood, toward more innovative approaches involving development of skills and microenterprises and the use of information technology. Comanagement and community-based natural resource management (CBNRM) strategies can provide a framework for such linked development and management initiatives. Community-centred comanagement can serve as a mechanism not only for resource management but also for social, community, and economic development by promoting participation and empowerment of people to solve problems and address community needs (see Chapter 8).

6.4 Enforcement and compliance

The inability to enforce, in the field, regulations that make perfect sense in the meeting room has been the downfall of many fisheries, large and small-scale. Small-scale fisheries with large numbers of fishers widely dispersed in inaccessible places are particularly resistant to top-down enforcement. A host of factors come into play to make this type of enforcement ineffective. Small-scale fishers are among the poorest people in society. Therefore, the political and judicial will to enforce regulations on them is often absent, especially when the action is seen as taking food from the fishers' family. The fact that the impact may be short-term, and that there may be expectations of increased food availability in the long-term, is not persuasive in these situations. Furthermore, in most countries, the judicial systems are bogged down with cases that the courts inevitably perceive as more important than enforcement of fishery regulations.

In small-scale fisheries, enforcement is often closely linked with issues of rural development and unemployment. Given this and the considerations in the previous paragraph, we believe that most small-scale fisheries need a radically different approach to enforcement and compliance. This new approach is consistent with the thread woven throughout this book: stakeholder consensus and involvement in management. This is a lengthy process that requires new skills on the part of the manager, who must now, in addition to having technical capabilities, also be a mediator, facilitator, and educator. The assumption that underlies this approach is that when the stakeholders understand the problems and the benefits of taking action, and agree upon the actions to be taken, they will take part in the enforcement — at least to the extent of encouraging compliance.

6.4.1 Factors affecting enforcement and compliance

To highlight the theoretical and empirical dimensions of enforcement and compliance in three Asian countries, this section summarizes *Enforcement and Compliance with Fisheries Regulations in Malaysia, Indonesia and the Philippines* (Kuperan *et al.* 1997) and *Economics of Regulatory Compliance in the Fisheries of Indonesia, Malaysia and the Philippines* (Susilowati 1998).

6.4.1.1 The compliance problem

Fisheries are regulated to mitigate overexploitation and conflicts among user groups. Often, overfishing resulting from open access to the fishery is addressed with regulations that restrict gear and vessels, set minimum fish size limits, implement time and area closures and quotas, and require fishers to have licenses. User conflicts are often addressed with gear prohibitions or restrictions and zoning to separate user groups. Fishers, like most regulated economic agents, typically are controlled through monitoring, surveillance, and enforcement. Frequently, the most costly element of fisheries management programs is enforcement, which accounts for a quarter to over a half of all expenditures. Compliance with regulations is usually far from complete, seriously jeopardizing the effectiveness of management. This raises questions about whether there are ways to improve the cost-effectiveness of traditional enforcement and whether there are ways to secure compliance without heavy reliance on costly enforcement.

Most modern analysis of compliance behaviour centres on deterring rational individuals from violating rules. Because individuals pursuing self-interest can harm others, it has been argued that social harmony can be realized only by controlling aspects of human nature. The basic deterrence model assumes that the threat of sanctions is the only policy mechanism available to improve compliance with regulations.

This deterrence model, however, has at least two important shortcomings. First, it does not explain the available evidence very well. Second, the policy prescriptions are impractical. The model assumes self-interested individuals weigh the potential illegal gains against severity and certainty of sanctions when deciding whether to comply. If the gains from illegal fishing are greater than those from legal fishing, the expected penalty should be large enough to offset the difference between legal and illegal gains. Since enforcement is costly, the probability of detection and conviction should be kept low and penalties high. The probability is usually low in practice. The typical odds of being caught violating a fishery regulation are below one percent. Penalties, on the other hand, generally are not large relative to illegal gains.

Raising penalties to the point where the expected penalty offsets illegal gains generally is not feasible. The courts are not willing to mete out sanctions that fit the crime, as measured by the illegal gains realized or the social harm caused by the detected and proven violation. The basic deterrence model predicts that the generally modest sanctions will not be an adequate deterrent to illegal fishing. Despite this apparent weakness, however, most fishers normally comply with regulations. Data show that 34 percent, 81 percent and 30 percent of fishers in Malaysia, Indonesia, and the Philippines, respectively, comply with the zoning regulation.

When asked why they comply when illegal gains are much larger than the expected penalties, many fishers expressed a sense of obligation to obey a set of rules. This moral obligation may be a significant motivation that explains much of the evidence on compliance behaviour. Other factors determining compliance are severity and certainty of sanctions, individuals' perceptions of the fairness and appropriateness of the law and its institutions, and social environmental factors. Compliance is linked to both the internal capacities of the individual and external influences of his or her environment. A fishery's law enforcement activity, by and large, determines the extent of compliance with its laws and regulations. Compliance is directly related to the effectiveness of fisheries enforcement. Enforcement is necessary to achieve the goals of fisheries management, but enforcement is costly.

In summary, the literature identifies the following factors that determine compliance: potential illegal gain, severity and certainty of sanctions, an individual's moral development and his or her standard of personal morality, an individual's perception of how just and moral are the rules being enforced, and social environment.

6.4.1.2 A Southeast Asian perspective on enforcement and compliance

The study by Kuperan *et al.* (1997) tested a model of compliance behaviour in which rational individuals are driven by internal and external motivations. The model accounts for morality, legitimacy and social influence in addition to the conventional costs and revenues associated with illegal behaviour. The study examined non-compliance behaviour of fishers in Indonesia, Malaysia, and the Philippines.

The results from the study provided some support for traditional enforcement policy. A higher rate of detection and conviction arising out of enforcement activities has the potential to discourage people from committing illegal activities. Similarly, more expensive penalties and fines imposed will make fishers comply with rules or regulations. In practice, however, probability of detection is low and violations are rarely detected, especially in Indonesia and the Philippines, given their geographic area and limited resources. According to the theory, levels of compliance can be improved by increasing the probability of detection and conviction or penalty rate. However, this course is not very practical because of the large financial requirements needed to attain such goals. With this in view, it is recommended that governments enhance enforcement resources and increase the penalty rate to deter violators.

According to the compliance theory, the willingness to comply that stems from moral obligations and social influence is based on the perceived legitimacy of the authorities charged with implementing the regulations. Other evidence suggests that a key determinant of perceived legitimacy is the fairness built into the procedures used to develop and implement regulatory policy. To the extent that this is valid, enforcement authorities should determine what policies and practices are judged fair by segments of the population subject to regulation. This may mean that civil penalties and other sanctions should be comparable in value to the larger harm done or gains realized. Therefore, policymakers and enforcement authorities would need to reveal to violating fishers, and to society at large, the extent of damage the violations caused so that they understand the procedural and outcomes (justice) aspect of the law and the penalties.

Although legitimacy of the management measures was seen as one of the determinants of compliance, the study results did not unanimously support the theory. The difficulty in understanding the concept of legitimacy may have contributed to the weaker performance of legitimacy variables. Another possible reason for the poor performance of the legitimacy variable is that other factors, not captured in the model but important enough to influence the normative factors of legitimacy, were overlooked (for example, institutional problems and enforcement weakness).

Overall, the results show that basic deterrence, moral development and social standing variables in all models are statistically significant in determining the violation behaviour of fishers in the selected study area. The legitimacy variables were not all significant. The study found sufficient support to demonstrate that personal moral development plays a more important role than legitimacy in securing compliance. This conclusion is consistent with Tyler's finding (1990) that process variables play more important roles than outcome variables.

6.4.1.3 Implications for comanagement

These results have important implications for comanagement. It is often stated that fisheries comanagement is likely to receive greater support from fishing communities when the communities are closely involved in the process and in determining the outcome. The findings of the study support this view. In a comanaged fishery, there is a greater moral obligation on individuals to comply with rules and regulations, since the fishers themselves are involved in formulating, rationalizing, and imposing the rules and regulations for their overall well-being.

6.4.2 Self-enforcement by fishers

Enforcement of regulations by the stakeholders themselves is increasingly being considered by governments short of enforcement resources. Stakeholder enforcement can take two forms that are not mutually exclusive. In the first form, fishers perform a mainly monitoring function, reporting violations to the authorities and exerting peer pressure. The sea urchin fishers of Barbados chose this option (**Box 3.2**), agreeing that they would be prepared to call a police hot line if there was a commitment by police to respond.

In the second form, fishers or other community members are designated as enforcement officers. Stakeholders in the Portland Bight Management Area, Jamaica, chose this option (CCAMF 1998). The Marine Conservation Project for the island of San Salvador, Philippines, also adopted the latter approach of a stakeholder enforcement program (Katon *et al.* 1997). Monitoring and enforcement of the marine sanctuary became the combined responsibility of the village police (three persons), a specially formed group called the "Guardians of the Sea" (five persons) and eight volunteers. Once equipped with boats and radios, these guards were able to apprehend individuals for a variety of violations (39 violations in the first eight months). The majority of violators were non-residents.

6.5 Conclusions

In this chapter we examined various aspects of implementing management for small-scale fisheries. Target Reference Points, as well as the more recent concept of Limit Reference Points, can apply to these fisheries but those derived from conventional population models may require more information than is commonly available or cost-effective due to the total value of the fishery. Therefore, to improve a fishery, we suggest targets and limits based on simpler approaches and agreed upon by the stakeholders. Consensus can be crucial when the actions cannot be derived from an objective algorithmic process. We also suggest the concept of Management Reference Directions. We emphasize the role that consensus among stakeholders can play in determining the desired direction and the actions that will take the fishery in that direction. Thus, tools for reviewing a variety of information and reaching group consensus must be added to the manager's repertoire of skills.

New targets, limits, and directions related to ecosystem health and functioning should feature in the management of small-scale fisheries. This is particularly true of inland and coastal fisheries, which are most vulnerable to non-fishery activities. Therefore, managers and stakeholders may need, at times, to put more effort into preventing and mitigating these influences than into controlling the fishery.

It seems to us that the control measures available to the small-scale fishery manager will be much the same as those available to the conventional fishery manager, and, indeed, as have been traditionally used by societies around the world. However, for small-scale fisheries, measures that are easily communicated, that intuitively relate to the status of the resource, and that are enforceable at the community level, will probably be more successful.

Chapter 7
Managing the Commons

7.1 Introduction

Although most conventional fishery management concerns itself mainly with biological or bio-economic methods, fishery management is really "people management," as recognized by generations of fishery managers (Gulland 1974). Paramount among people management problems is the "tragedy of the commons," which is about the divergence between individual and collective rationality. Resources that start out abundant and freely available to all tend to become ecologically scarce. Unless their use is somehow regulated in the common interest, the long-term outcome will be ecological ruin for all (Hardin 1968). The purpose of the chapter is to explain the property rights approach, and the implications of this approach to fisheries management.

The chapter begins with a consideration of the "tragedy of the commons" and its solutions, since the issues raised are fundamental to fisheries management (Section 2). A section on the question of who makes the rules and regulations (Section 3) follows the discussion of theory and practice in solving the "tragedy". However, fisheries management consists of more than just rules and regulations; management must also aim at building and maintaining fishing communities that can make their own rules and solve their own problems (Section 4). Thus, involving fisherfolk in managing the fishery raises the issue of their capability to contribute to management; these are the questions of capacity building and institution building (Section 5).

7.2 "Tragedy of the Commons" and its Solutions

In Garrett Hardin's "tragedy of the commons" story, a group of medieval English herders keep increasing the size of their individual herds, eventually exceeding the carrying capacity of the village commons and losing all (**Box 7.1** and **Figure 7.1**). At the time Hardin's paper was published, many people found the "tragedy" metaphor insightful and applicable to world's fisheries. The earliest theories on the commons had, in fact, been based on fisheries. Indeed, the problem of common-property resources is also known as the "fishermen's problem" (McEvoy 1986). Fisheries provide the ultimate example of the commons dilemma: the resource is fugitive, and the fish you do not catch today may be caught by someone else tomorrow. It is difficult to see a fisher's incentive to conserve the resource, as opposed to catching as much as possible, as soon as possible. But since each fisher operates with the same rationality, the users of the fishery commons are caught in an inevitable process that leads to the destruction of the very resource on which they all depend. Because each user ignores the costs imposed on others, individually rational decisions accumulate to result in a socially irrational outcome.

Box 7.1 The "tragedy of the commons"

As originally formulated by Garrett Hardin, the "tragedy" is the outcome of economically rational, individualistic decision-making. "Picture a pasture open to all," said Hardin. Each cattle owner will want to maximize gains by keeping as many cattle as possible. But sooner or later, the carrying capacity of the land will be reached. Explicitly or implicitly, each herder will ask, "What is the utility to *me* of adding one more animal to my herd?" Each new animal will bring the herder a positive utility of nearly +1. But the effects of overgrazing will be shared by all, and the herder's loss will only be a fraction of -1. Thus, the herder's rational decision, said Hardin, "is to add another animal to his herd. And another and another ... But this is the conclusion reached by each and every rational herdsman sharing the commons. Therein is the tragedy."

If the only commons of importance were a few grazing areas, the "tragedy" would be of little general interest. But almost all resources, including all fisheries except aquaculture, can be considered a commons jointly used by many, in which potential users are difficult to exclude, and the activity of any one user may affect the welfare of all others. Hardin himself used the grazing commons as a metaphor for the problem of overpopulation. The dominant legacy of his famous essay, however, has been in the area of natural resource management. The phrase, "tragedy of the commons" has stuck, even though many scholars have noted that commons operated successfully for several hundred years in medieval England, and have questioned if a tragedy of the sort described by Hardin ever occurred widely.

The best-known formulation of the commons dilemma is Hardin's, but the history of the concept probably goes at least as far back in time as Aristotle, who observed, "what is common to the greatest number has the least care bestowed upon it. Everyone thinks chiefly of his own, hardly at all of the common interest." Two fishery resource economists, Gordon (1954) and Scott (1955), are usually credited with the first statement of the theory of the commons — more than a decade before Hardin. By the time Hardin's famous essay appeared, fishery economists were already modelling the attraction of excess labour into the fishery and the resource depletion that followed. What did Hardin's "tragedy of the commons" contribute to fisheries management?

Hardin's (1968) proposed solutions were either top-down management by the state or the privatism of free enterprise. Many governments used the "tragedy of the commons" analysis to rationalize central government control of all kinds of common resources. For example, Matthews (1988) has shown that the "tragedy" analysis shaped policy in the fisheries of Atlantic Canada, playing a central role in government interventions and privatization. The "tragedy" analysis leads to a pessimistic, disempowering vision of resource management. Users are seen as trapped into a situation they cannot change. It is therefore argued that solutions must be imposed on the users by an external authority.

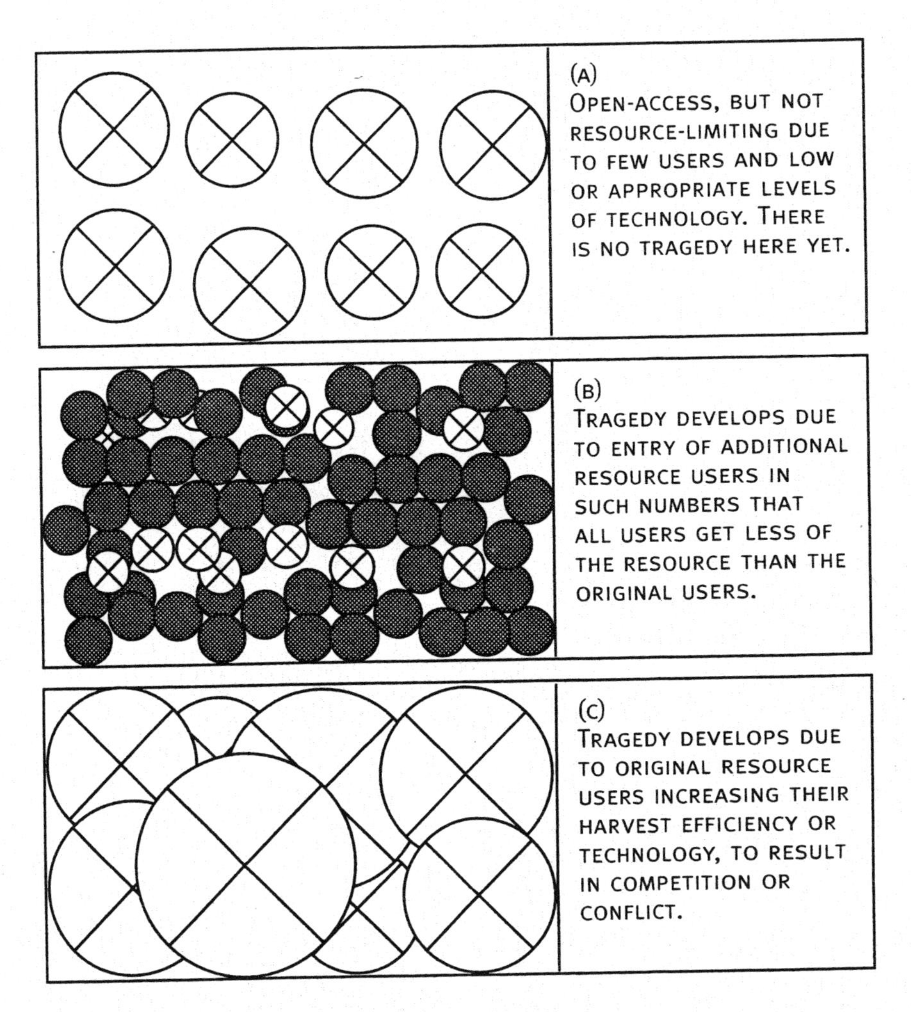

Figure 7.1 Illustrations of the tragedy of the commons.

Even though some of the best-known examples of the "tragedy of the commons" come from the area of fisheries, it is also obvious that, for thousands of years, resource users have organized themselves to manage common resources such as fisheries, and have developed and maintained institutions to govern these resources (Ostrom *et al.* 1999). Findings from a large number of cases covering a diversity of resource types, geographical areas, and cultures have revealed the existence of local and traditional management systems and of commons institutions (McCay and Acheson 1987; Berkes 1989; Bromley 1992). These institutions — that is, local norms and rules — have been found to exist even in the absence of any government regulations (**Box 7.2**). These findings, mainly in the last three decades, have required a re-assessment of the "tragedy of the commons" as the correct explanation of the human use of fisheries and other commons.

BOX 7.2 DEFINING INSTITUTIONS AND COMMON-PROPERTY (COMMON-POOL) RESOURCES

Institutions are defined as "humanly devised constraints that structure human interaction. They are made up of formal constraints (rules, laws, constitutions), informal constraints (norms of behaviour, conventions, and self-imposed codes of conduct), and their enforcement characteristics" (North 1993). Institutions are "the set of rules actually used (the working rules or rules-in-use) by a set of individuals to organize repetitive activities that produce outcomes affecting those individuals and potentially affecting others" (Ostrom 1992). It is also important to note that institutions are socially constructed; they have normative and cognitive, as well as regulative, dimensions (Jentoft *et al.* 1998). The cognitive dimension has to do with the questions of the nature of knowledge and the legitimacy of different kinds of knowledge, relevant to the use of traditional ecological knowledge in fisheries management

We define *property* as the rights and obligations of individuals or groups to use the resource base; a bundle of entitlements defining owner's rights, duties, and responsibilities for the use of the resource, or "a claim to a benefit (or income) stream, and a property right is a claim to a benefit stream that some higher body — usually the state — will agree to protect through the assignment of duty to others who may covet, or somehow interfere with, the benefit stream" (Bromley 1992). *Common-property* (*common-pool*) resources are defined as a class of resources for which exclusion is difficult and joint use involves subtractability (Berkes 1989; Feeny *et al.* 1990). Institutions have to deal with the two fundamental management problems that arise from the two basic characteristics of all such resources: (1) How to control access to the resource, given that it is difficult or costly to exclude potential users from gaining access to the resource (the exclusion problem), and (2) how to institute rules among users to solve the potential divergence between individual and collective rationality, that is, how to deal with the problem that each person's use of the resource subtracts from the welfare of the others (the subtractability problem).

The fundamental flaw in Hardin's "tragedy" is the assumption that users could freely and openly access a common resource. Thus, Hardin's metaphor is misleading for policymakers and resource managers because it confused "common property" with "open-access." Resource economists Ciriacy-Wantrup and Bishop (1975) were among the first to point out that "common property" is not the same as "everybody's property." Common property refers to a class of property rights, usually a right to use something in common with others and a right not to be excluded from its use. Typically, common property involves "a distribution of property rights in resources in which a number of owners are co-equal in their rights to use the resource" (Ciriacy-Wantrup and Bishop 1975). By contrast, open-access is *laissez-faire* or a free-for-all, a condition of no property rights at all.

A second flaw in the "tragedy" idea is that it ignores the social relations that characterize resource users throughout the world. Hardin's analysis, in common with that of many neoclassical economists, would have us believe that resource users are self-centered utility maximizers, unrestricted by community and social relations. As anyone who has worked with fisherfolk knows, even the most selfish and individualistic fishers are nevertheless subject to social pressures that shape their behaviour. Social scientists were the first to be skeptical of the "tragedy" analysis because they tend to be familiar with "... the social and moral aspects of user behaviour. Users form communities. Natural resource extraction is guided by social values and norms, many of them non-contractual, some of which stress moderation and prudence" (McCay and Jentoft 1998).

7.2.1 Property-rights regimes

Much of the academic literature on commons since Hardin's seminal idea has concentrated on exploring the potential solutions to the tragedy of the commons, and narrowed it down to basically three property-rights regimes:

- **State property**, with sole government jurisdiction and centralized regulatory controls;
- **Private property**, with privatization of rights through the establishment of individual or company-held resource harvesting quotas (such as ITQs);
- **Communal property**, in which the resource is controlled by an identifiable community of users, and regulations are made and enforced locally.

These three categories, along with **open-access** as a property-rights regime, make up four ideal analytical types of property rights. In reality, many resources are held under regimes that combine the characteristics of two or more of these types. State property regimes may employ market mechanisms (such as quotas) and locally enforced social mechanisms (such as self-policing). The use of quota controls may include centralized regulatory controls and local-level monitoring and enforcement of quotas. Common property regimes may include enforcement backup by government controls. All three regimes — communal property, private property and state property — involve defined property rights, whereas open-access is the *absence* of property-rights. It is open-access that results in the "tragedy of the commons." There is nothing inherent in the commons that leads to a "tragedy."

Figure 7.2 illustrates a hypothetical coastal area in which several property-rights regimes are found together. On the coast is a private aquaculture area, next to a fishing territory controlled by a village. The coastal fishery within the territorial sea (up to 12 miles) and the offshore fishery in the Exclusive Economic Zone (EEZ usually up to 200 miles) of that nation, is under state property. Beyond the EEZ, there may be an international regime in force on the high seas, or the fishery may really be open-access. It may be open-access within the territorial sea and EEZ as well, if the state is unable to make or enforce its regulations. The private and the communal areas may both be mixed regimes. It is the state that normally leases aquaculture areas, and it is also the state that enables a community to control its fishing area, as in comanagement (Chapter 7). Although the example is hypothetical, many coastal areas do have co-existing and overlapping property-rights regimes. Resource managers cannot function effectively unless they know the property-rights regimes they are dealing with, and the implications of each with respect to solving the "tragedy of the commons."

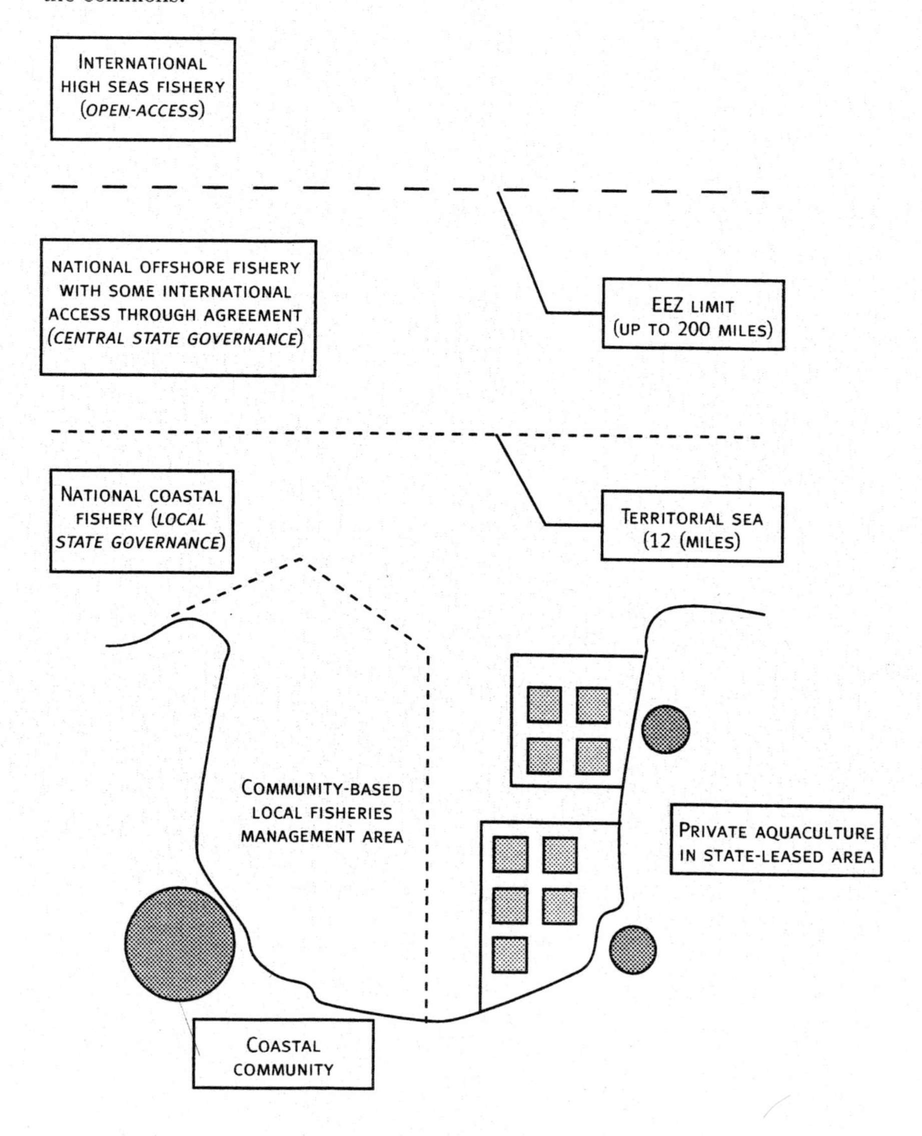

Figure 7.2 Property and fisheries management regimes.

7.3 Who Makes the Rules to Solve the Commons Dilemma?

Solving the "tragedy" starts by addressing the two basic characteristics of the commons, the exclusion problem and the subtractability problem. That is, how to control access to the resource (the exclusion problem), and how to make, and enforce rules and regulations among users to reduce their impact on one another (the subtractability problem).

Controlling access to the resource is a common problem. In many parts of the world, fishery managers do not possess good inventories at their disposal regarding the users of a given resource: Who are the user groups? How many boats and fishers are there of different kinds of users? In some cases, coastal rural populations are very fluid; participants in a fishery are changing all the time. In other cases, a nation's legislation defines the sea as open-access; the manager has no means (or desire) to enforce access control or comanage access with communities that have traditional resource rights in an area. But the findings of the common-property literature are clear on this issue. If access to a resource is not controlled by some means, sooner or later the resource will be subject to a "tragedy of the commons."

The subtractability problem is also common, and manifests itself through resource use conflicts among fishers or through non-compliance with regulations. When fishermen violate regulations, the managers must ask themselves: Do the regulations make sense, and are they serving the management purpose for which they were intended? Can fishers help make and enforce better regulations? Is the government helping fisherfolk organizations to make and enforce their own rules or hindering them? Findings of the common-property literature are equally firm on the question of subtractability: management does not work unless there is a set of rules that all the users agree to follow.

The rules to control access and address the subtractability problem may be made by the government, by the market, by the fishing communities themselves, or by any combination thereof. Each fishery is unique and there are no set solutions. Managers need to know not only the characteristics of their fisheries but also something about the international experience. This section presents the strengths and limitations of each of the three property-rights regimes — state property, private property, and communal property—to solve the exclusion (access) and the subtractability problems.

7.3.1 Role of the state: Limits to government management

Regarding the control of access to the resource, the state property regime is most effective in situations in which the general public good is involved, and where property rights regimes are not suitable for solving the exclusion problem. For example, a marine protected area cannot be managed under a private property regime (i.e. individual or company property), or a communal property regime favouring only one group, if the area needs to serve many stakeholders or the general public good. But what about marine fish stocks in general? Almost all western industrialized countries now use government powers to limit the access of the general population to the fishery resource. This is usually accomplished by limiting the number of fishers or fishing licenses. Among developing countries, however, licensing of fishers is common, but limiting the numbers to control access, is not.

The reasons for this vary. Many coastal rural people do some fishing on a part-time or seasonal basis, and this is considered a natural right. In any case, many developing country governments encourage their citizens to be productive, and provide incentives for the fishery sector to grow and produce more. Such policies do not necessarily result in overfishing if this fishing is low-technology and low-intensity. Scale of the fishery is important. But if the fishery is really open-access, resources will be depleted sooner or later, according to the theory.

However, when high-technology fishing units enter an open-access fishery, resource depletion is usually rapid. This has happened in many Pacific island fisheries (Johannes 1998a). For this reason, some developing countries are beginning to use state-level controls to limit licences in the high-technology, mid- to large-scale sector, but without attempting to impose access controls on the small-scale inshore sector, relying instead on the self-regulation of local fishing communities.

There is another important limitation of state management in many developing countries. For most resources, exclusion and subtractability problems are not necessarily solved by declaring the resource to be state property and simply passing a set of regulations. In many countries, management infrastructure is inadequate, enforcement lacking, and budgets insufficient. The protection of a national resource may be dependent on the ability and will of local groups to support a given government measure. For example, in the Caribbean island state of St. Lucia, the protection of marine resources through the establishment of Maria Islands Marine Reserve was successful only when the local community supported and helped enforce the boundaries (Smith and Berkes 1991).

Regarding the subtractability problem, the lack of resources to manage and enforce government regulations has always been a problem. In high-income countries and low-income countries alike, there is no lack of government regulations. However, countries vary greatly in the enforcement of regulations. Very few countries and fisheries departments are able to enforce all of their regulations. The inability to enforce even such apparently simple rules such as the ban on dynamite fishing, has forced central governments to look for alternative solutions. In the Philippines, for example, decentralization of fishery management through the *Local Government Code* and the *Fisheries Code* of 1998 is interpreted in part as a the failure of the central government to regulate coastal fisheries, and to pass on the responsibility to municipal governments and local communities (**Box 7.3**).

Box 7.3 Philippines municipal fisheries

Philippines coastal fisheries have been characterized by intensive, competitive exploitation, conflicts between gear groups, resource depletion, and enforcement problems. The government has instituted policies favouring decentralization, perhaps due to the lack of government management and enforcement capability.

The Philippines government in 1991 enacted into law the *Local Government Code* (LGC) which sought to decentralize government functions and operations to local governments. The LGC granted local governments (municipalities) a number of powers, including the management of municipal or near shore waters, defined as all waters within 15 km of the coastline. The general operating principle is that local government units (LGU) may group themselves, consolidate or coordinate their efforts, services, and resources for a common purpose. The LGUs and local communities were also given certain privileges and/or preferential rights. For example, municipalities were given the exclusive authority to grant fishery privileges in municipal waters and to impose rentals, fees, and charges. In terms of fishery rights, the organizations or cooperatives of marginal fishers were granted preferential rights to fishing privileges within the municipal waters, such as the erection of fish corrals and gathering fish fry.

The decentralization of the Philippines fisheries is consistent with three decades of experience with citizen participation, NGO involvement, and comanagement, starting with forestry and water resources. Section 35 of LGC specifically states that LGUs may enter into joint ventures and such other cooperative arrangements with peoples' organizations and non-governmental organizations to engage in the delivery of certain basic services, projects for capacity building and livelihood security, and to develop local enterprises designed to diversify fisheries.

Source: Pomeroy (1994b); Pomeroy and Pido (1995).

7.3.2 Role of market controls: Limits to private property regimes

In all but a few countries, private property rights have provided a solution for the exclusion problem in the use of agricultural land resources, but not for common-property resources such as fisheries, wildlife, forests, and grazing lands. This is in part because agricultural land is relatively easy to delineate and defend; a fishing area or a forest is not. A second reason is the "divisibility" of the resource. Farmland is relatively easy to divide into discrete lots; fishing areas are not. Some success has been achieved in privatizing seaweed cultivation plots in Zanzibar, leased from the government to a company and then divided into individual plots. Similarly, aquaculture ponds are in discrete areas and can be privatized readily; wild fish cannot. With many types of commons, private property rights do not provide an appropriate mechanism for solving the exclusion problem. An alternative is the privatization of *harvesting rights*.

Many Western industrialized countries have been working for several years with the privatization of quantitative harvesting rights. The system of individual transferable quotas (ITQs), regulated by the government, enables market forces to direct the allocation of resources, hence increasing economic efficiency. Under an ITQ system, each quota-holder receives a share of the total allowable catch for a given species, and these quotas can be bought, sold or leased. ITQ have been used in large-scale fisheries in which the numbers of enterprises are small. They have the potential to work effectively *if* the total allowable catch can be reliably estimated species by species, harvests monitored, and the bycatch problem minimized. Harvest monitoring, in particular, appears to be crucial. Quotas do not work well when the resource is used by many coastal communities (rather than a few large companies), when the total allowable catch cannot be forecast, where there is a mixed fishery and an incidental catch problem, and when fishing units are small and enforcing quotas not feasible (Wilson *et al.* 1994).

Regarding the subtractability problem, privatization can be efficient, it is argued, because it reduces the extent of government regulations. Privatization provides incentives for the owner to regulate resource use. The theory is that if the owner has property rights in the resource and those rights can be traded both the costs and benefits of resource use will accrue to the same owner. This would eliminate the divergence between individual and collective interests, thereby solving the "tragedy of the commons." The costs and benefits will be reflected in the market price of the resource, giving the owner the incentive to regulate resource use in a manner consistent with private objectives.

These incentives may be consistent with private economic efficiency, but they are not necessarily consistent with resource conservation. Clark (1973, 1976) pointed out that, whether incentives created by privatization are consistent with sustainability will depend on the biological characteristics of the resource and the economic characteristics of the market. In general, Clark showed that for slow-growing and late-maturing species, it may be economically optimal to deplete the resource rather than to use it sustainably. Private property rights permit the owner to regulate his/her own use to maximize the present value. This gives the owner the option of seeking higher returns by mining out the resource (rather than using it sustainably) and re-investing the capital elsewhere. It is in fact the experience in many parts of the world that mobile fishing fleets will not use a local resource sustainably but fish it out and move on.

7.3.3 Role of social controls: Limits to community-based management

In contrast to industrial large-scale fisheries, community-based small-scale fisheries have much stronger incentives to use a local resource sustainably. Small-scale fishers have fewer options for geographic or occupational mobility. Their families are also dependent on local resources, leading to the development of communal property regimes. Under communal property regimes, exclusion means the ability to exclude people other than the members of a defined group. One of the lessons in the early common property literature was that the legal recognition of communal sea tenure could lead to sustainable resource use. Christy (1982) referred to these rights as Territorial Use Rights in Fisheries or TURFs. The logic behind the TURFs approach is that the people who have the most to win and lose from the management of the fishery, the local community, can regulate themselves if the community's fishing territory can be delineated and the fishery in it can be monitored and controlled by the community. A number of Pacific island states have moved in the direction of formally recognizing such traditional marine tenure systems (Baines 1989; Ruddle 1993). The recognition of TURFs is not limited to small island states. A well-known and documented example of successful exclusion under legally recognized communal resource use area is the Japanese coastal fishery system (Ruddle 1987). As well, many experiments are in progress in various parts of the world. The evolution of successful commons management through the legal recognition of local use areas has been documented in a number of cases, including mangrove forests in St. Lucia in the Caribbean (Smith and Berkes 1993).

In many parts of the world, however, there is no legal recognition of exclusion under communal property regimes. In such situations, the exclusion of outsiders by the local users has been informally enforced through such means as local customs, social sanctions, threats, and even violence. A case in point is the *sasi* system in the Maluku Islands of eastern Indonesia (Harkes and Novaczek 2000). Community exclusion of outsiders occurs in some fisheries in the West as well. The lobster fishing territory system in Maine, USA, was not recognized until recently by the government, and was technically illegal. But these territories have been judged to function well and contribute to the management of the resource (Acheson 1988), and the State of Maine has recently begun to use them in its zoning.

The Maine lobster case is merely one of many (see examples in Berkes 1989; Bromley 1992), but it is a significant example because it comes from a country in which individual rights and free-access to marine resources are deeply held cultural beliefs. Elsewhere, reef and lagoon territorial systems are most commonly found in the Asia-Pacific and in coastal areas of many developing countries. Even in post-colonial countries where the original inhabitants and their traditions have all but disappeared, rudimentary systems of marine tenure seem to have evolved over the shorter period of recent settlement. One such system, on the north coast of Jamaica, lacked the sophistication of those from Oceania (Johannes 1978) but was real enough to limit the access of outsiders to coral reef fish resources (Berkes 1987).

7.3.4 Can fishing communities make their own rules?

Regarding the solution of the subtractability problem, a number of cases and several books document that fishing societies are capable of making their own rules to manage resources on which their livelihoods depend (Cordell 1989; McGoodwin 1990; Dyer and McGoodwin 1994). Wilson *et al.* (1994) surveyed 32 detailed studies of local-level regulation in small-scale or traditional societies from all over the world, and found that rules almost always focused on fisher behaviour and qualitative controls, and not on quantitative controls such as quotas. That is, as a general pattern, traditional systems are characterized by rules and practices that seek to regulate *how* fishing is done, as opposed to focusing on the amount harvested, as in ITQs (see Chapter 6).

In the order of decreasing importance, these societies made and used rules pertaining to territorial controls, access limits, seasonal limits, technology restrictions, breeding stock protection, protection of juveniles, and size limits (Wilson *et al.* 1994). **Table 7.1** shows in more detail the kinds and variety of traditional rules in place in one part of the world, the tropical Pacific. The table is summarized from the work of Johannes (1978, p. 352), who points out that "almost every basic fisheries conservation measure devised in the West was in use in the tropical Pacific centuries ago." These findings are significant: "The fact that such regulations are found so widely and have lasted for such a long time suggests that such rules were highly adaptive" (Wilson *et al.* 1994, p. 305).

Table 7.1 Marine conservation measures employed traditionally by tropical Pacific islanders.

Method or Regulation	Examples
Closed fishing areas	Pukapuka; Marquesas; Truk; Tahiti; Satawal
Closed seasons	Hawaii; Tahiti; Palau; Tonga; Tokelaus
Allowing a portion of the catch to escape	Tonga; Micronesia; Hawaii; Enewetak
Holding excess catch in enclosures	Pukapuka; Tuamotus; Marshall Islands; Palau
Ban on taking small individuals	Pukapuka (crabs); Palau (giant clams)
Restricting some fisheries for emergency	Nauru; Palau; Gilbert Islands; Pukapuka
Restricting harvest of seabirds and/or eggs	Tobi; Pukapuka; Enewetak
Restricting number of fish traps	Woleai
Ban on taking nesting turtles and/or eggs	Tobi; New Hebrides; Gilbert Islands
Ban on disturbing turtle nesting habitat	Samoa

Source: Adapted and summarized from Johannes (1978).

Not all examples of successful rule making are historic or based on tradition. A study of Turkish Mediterranean coastal fisheries found that rules for fishing site allocation and conflict-reduction were developed in the 1970s and 1980s. They were based on meetings of the fishermen in the local teahouse. Initial allocations of fishing sites were by done by lottery and the sites were rotated daily after that until the end of the fishing season. The rotational use of some 37 sites reduced conflict and provided equitable opportunity for all group members, about 100, to the prime netting areas of which there were four (Berkes 1986). These rules for self-governance were found to have evolved over a period of one decade. The design rules (lottery and rotation) did not solve the problem of increasing numbers of boats through the 1980s, but formed the basis for the diversification of fishers into the developing tourism industry (Berkes 1992). Similar local management systems with allocation rules have been documented from several areas. A particularly sophisticated and apparently effective local system from Sri Lanka is summarized in **Box 7.4**.

Box 7.4 Kattudel *lagoon fisheries of* Sri Lanka

Local governance systems and allocation rules for lagoon resources have developed in a number of different geographical regions. One such contemporary lagoon management system has been described from the Negombo estuary in western Sri Lanka. The case study illustrates the sophisticated level of governance that can be achieved by traditional systems and the key role of local institutions. Of several kinds of fishing operations carried out in the Negombo lagoon, the one known as the *kattudel* fishery (which uses a kind of trap net) targets high-value shrimp, *Metapaneus dobsoni*, as the shrimp migrates out to the sea. At the mouth of the lagoon, there are 22 named fishing sites at which 65 nets could be used at any one time.

Members of a defined group whose fishing rights go back at least to the 18th century exclusively use the sites. The sites are presently controlled by members of four Rural Fisheries Societies (RFS) based in the villages around the lagoon. Elaborate rules govern eligibility and membership in the RFS, the obligations of the fishers, and the system by which the four RFS cooperatives share the resource, and the allocation rules within each RFS. There are about 300 members in the *kattudel* fishery out of some 3 000 fishers using the lagoon. The members take turns at the 22 fishing sites. A lottery system is used to allocate turns (one night at a time) and produce a rotation through all 22 sites, to give each fisher an equitable opportunity.

One lesson of the *kattudel* fishery is that the rules of the fishery, most recently reorganized in 1958, have legal status under Sri Lanka's *Fisheries Ordinance* as "Negombo (Kattudel) Fishing Regulations." This is significant because only through legal enforcement can the strict limits on membership, and hence the limited-access nature of the fishery, be maintained. A second lesson of the *kattudel* case is that lagoon fisheries and associated traditional management systems can be sustainable over long periods. But they are not necessarily stable over time. The *kattudel* fishery went through turbulent times, most recently in the 1940s and 1950s, and evidently survived the various crises. The real test of a management system is not whether it has "perfect rules" but whether rules can be successfully redesigned or adjusted in response to crises.

Source: Atapattu (1987); Amarasinghe et al. (1997)

7.3.4 WHAT DETERMINES SUCCESS OR FAILURE?

The literature contains many examples of successful community-based fishery systems but also documents two general kinds of failure. The first is "community failure" or the inability of the group to regulate its own affairs (McCay and Jentoft 1998). The second is related to the impact of external forces on local institutions, including technology change, economic change and population pressure (Berkes 1985; Ruddle 1993). The effects of the creation of open-access conditions by external forces, such as colonialism, have been documented particularly well in Oceania. Local-level control mechanisms, seen as "too conservative" by colonial administrators, were eliminated to open up trade, with the result that a suite of marine resources were serially depleted, area after area, in the Asia-Pacific region (Johannes 1978).

For communal systems to work well, a number of conditions have to be satisfied. These key conditions for successful management are dealt with in more detail in Chapter 8 on comanagement. As discussed further in Chapter 8, the challenge is how to reconcile local-level rules and government regulation toward improving fishery management. In traditional societies, resource use rules were made and enforced locally. In modern societies, such rules are usually made by central governments. The ultimate responsibility for the resource rests with the government. But it may be more efficient and effective for government resource managers to share management powers and responsibility with fishing communities, and to have some of the rules made and enforced locally (Pomeroy and Berkes 1997). That is in fact the essence of the management system used in Japanese inshore fisheries. It recognizes the importance of the local decision-making and enforcement, by devolving certain kinds of management authority from the central government to the fishing community (**Box 7.5**). Such management recognizes the central importance of self-regulatory abilities of coastal communities, as part of the biological and economic objectives of fish harvesting.

7.3.5 THE COLLECTIVE ACTION PROBLEM

A number of lessons have been learned from the very large literature on the commons which has accumulated since the 1980s. One of the more fundamental findings is that common property regimes, as collective resource management systems, have been shown to develop when a group of individuals are highly dependent on a resource and when the resource is limiting (Ostrom 1990; Bromley 1992). Common property systems do not develop if the resource is superabundant. Repeated experiences with a resource problem, such as low catch or no catch, is often necessary for a response. However, the nature of the problem is such that, individual responses cannot solve the problem. For example, the resource that one fisher conserves today will probably be harvested by others. The solution will work only if all fishers agree to stick to rules that ensure tomorrow's catches.

BOX 7.5 JAPANESE COASTAL FISHERIES

The Japanese coastal fishery provides regulatory authority at national and regional levels, and decision-making power mainly at the local level. It implements management measures and solves disputes over resource use rights, and provides a legal safeguard of village-based resource rights. Until about 1900, these management functions in Japanese inshore fisheries were carried out by village guilds. The modern system was designed to formalize historical village fishing rights. With the implementation of the 1901 *Fisheries Law*, village sea territories that had evolved during the feudal era were mapped, codified and registered. Updated in 1949, the *Fisheries Law* gave rights and licences to working fishermen only, and placed fishery management in their hands through the local Fisheries Cooperative Associations (FCA).

Each FCA (or federation of FCAs) has exclusive ownership of coastal waters except port areas and industrial zones. FCAs apply to government for licences that they distribute among their members. Non-members cannot fish. Members who do not obey the rules are expelled. The FCAs control many aspects of the coastal fishing activity within their immediate jurisdiction by implementing and enforcing national fishery laws and regulations, supplemented or complemented by those made locally. For example, the national government establishes total allowable catch (TAC) for the offshore and coastal fishing areas. The allocation of the total quota to the various FCAs is done by the prefecture (regional government). The FCA then has the responsibility to allocate their quota among members. The FCA has close interaction with the national, prefecture, and municipal governments on matters including the design and implementation of management plans, approval of regulations, fishery projects, and budgets, subsidies, and licenses and other rights. FCAs also carry out marketing, processing, leasing fish equipment, purchasing supplies, and education functions.

The prevalent maritime tradition in Japan, unlike the West, never included the idea that the sea is (or should be) open-access. Instead, a complex system of locally varied marine tenure developed over many generations. Ownership of marine commons in coastal waters is quite comparable, in Japanese law, to the ownership of village commons.

Source: Ruddle (1987); Lim et al. (1995).

Hence, the principal problem faced by members of a group using a commons is how to organize themselves and to change from a situation of individual action to one of collective action. A collective action strategy is one that helps obtain greater joint benefits (e.g. making a livelihood) and reduces joint costs (i.e. resource depletion). Collective action occurs only if the group of commons users have the authority to make decisions and to establish rules over the use of the resource. But will the individuals stick to the rules made by the group?

There are always incentives for an individual to adopt opportunistic strategies and circumvent the rules for private gain. Free riders are those who benefit from the work of the group without contributing to that work. The challenge for any common property regime is to establish institutional arrangements that minimize transaction costs and counteract opportunistic behaviour such as free riding. Transaction costs are the costs of doing business. If the rules are clear and if everyone knows them, transaction costs will be low and rules relatively easy to monitor and to enforce (Feeney 1998).

Collective action entails coordination and organization problems that do not exist with other regimes, state property and private property. To maintain institutional arrangements over time, it is important to develop workable procedures for monitoring the behaviour of the resource users and to use social sanctions (or penalties) where necessary, and for settling conflicts. Again, transaction costs come into play for sanctions and for conflict management. Where a group of users know one another, have reciprocity, similar livelihood activities, and share similar values, it will be relatively easy to enforce sanctions and to manage conflicts.

By contrast, it would be relatively more difficult to do these things, because of high transaction costs, if the users do not know one another, do not reciprocate, do not make their livelihoods in a similar way, or do not hold similar values. The main reason that the commons literature refers so much to community-based resource management is that when the resource users are organized as a "community", this tends to lower transaction costs and increases the likelihood of successful organization towards collective action. Hence, the subject of community is an important one for a fishery manager.

7.4 Beyond Regulation: Managing Fishing Communities

The commons literature has established the key importance of the ability of a fishing community to manage its own affairs, sometimes requiring capacity building by the resource manager. Conversely, "community failure" limits the ability of the resource manager to bring fishing communities into the management process. There is a close relationship between communities and fish stocks, in traditional societies and in many modern nation states, small and large. Not only are viable fish stocks necessary for the vitality of fishing communities, but the reverse also holds true: "viable fish stocks require viable fishing communities" (Jentoft 2000).

Jentoft argues that communities that disintegrate socially are a threat to fish stocks. "Overfishing results when the norms of self-restraint, prudence and community solidarity have eroded. It occurs when fishermen do not care about the resource, their community and about each other. Then, their ability to communicate among themselves, to agree and cooperate is lost" (Jentoft 2000). What defines a viable fishing community? Jentoft does not provide guidelines to operationalize the concept, but the following are clearly important: the ability of members of a community to communicate among themselves; to be able to make rules, agree upon them, and enforce them; and to act collectively. **Box 7.6** provides a more comprehensive list of items, consistent with Jentoft, as used by Pomeroy *et al.* (1997), to measure local perceptions of community well-being, as well as that of the coastal ecosystem.

Noting that the role of "community" is important for the fishery manager who would like to deal with the fishery as a system, Jentoft observes that fisheries management

Box 7.6 Impact indicators used in measuring success of a development project in Philippines.

A community-based coastal resource management project was implemented in the Central Visayas, Philippines. In evaluating factors contributing to the success of the project, the researchers asked questions covering the following impact indicators:

1. Overall well-being of the household;
2. Overall well-being of the resource;
3. Local income;
4. Access to resources;
5. Control over resources;
6. Ability to participate in community affairs;
7. Ability to influence community affairs;
8. Community conflict;
9. Community compliance with resource management;
10. Amount of the traditionally harvested resource in the water.

The respondents were asked questions covering these items on ladder-like scale of 15 steps. They were asked for their view for the situation both before and after the project, so that the user perception of impacts, that is, their views on changes attributable to the project could be evaluated.

Source: Pomeroy et al. (1997)

in many countries is seen as a relationship between a government and a rights holder who is typically not a community or a household but an individual. The community is the missing element not only in Hardin's model of the "tragedy of the commons," but also in many government fisheries management systems around the world. Implicitly, fishers are perceived as competitors in the fishing commons. Their social relations are "positional," as in a bus queue; they do not have any relation other than being at a particular place and time, with the same goal in mind: to get on the bus and to find a good seat. From the individual point of view, all other passengers are competitors or a nuisance; they are in the way. In most fisheries management systems in the world, harvesters on the fishing ground are seen in this way.

An alternative image of fishers sees them as mutually dependent and supportive of one other, where individuals regard themselves as members of a social group, and where community is a system of reciprocal relationships. There is fisheries social science literature that describes local communities as learning systems and networks, emphasizing integrative social characteristics of communities (**Figure 7.3**). In this view, communities are not just aggregations of utility-maximizing individuals. Rather, "communities are well connected systems rooted in kinship, culture and history" (Jentoft 1999). This more optimistic view about cooperation in the use of fishery commons is supported by Prisoner's Dilemma models in the area of Game Theory. These models show that cooperation and reciprocity among users may evolve and prevent "tragedies of the commons" from occurring, even in the absence of an external authority (Axelrod 1984).

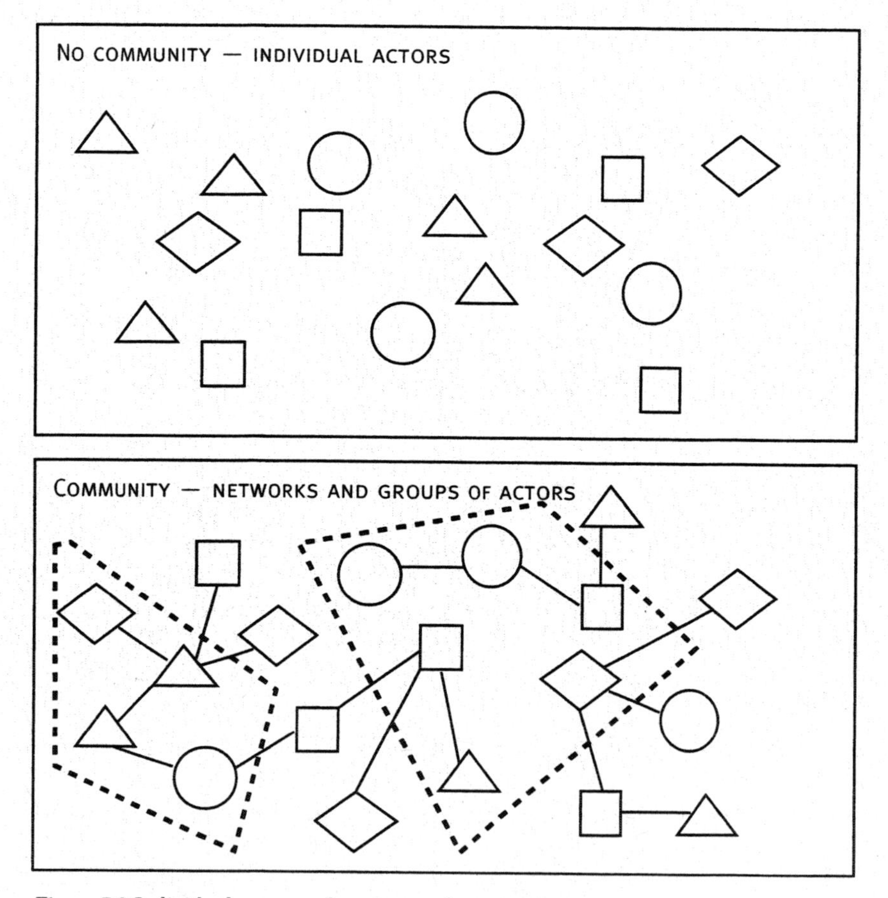

Figure 7.3 Individual actors and concepts of community.

7.4.1 Keeping "community" in perspective

However, one should be careful not to romanticize the fishing community. To be sure, one sees both cooperation and competition in most fishing communities, often in a dynamic tension. Community relations may include both symbiotic and destructive interactions. There are individual differences as well: utilitarian motives drive the behaviour of some fishers more than others. Communities are not homogenous. A community cannot be considered one uniform interest group. There are often gender, ethnic, and socioeconomic tensions within a community. Studies on the Atlantic Canada trawling industry have documented groupings of fishers within a community who have reciprocal exchange relations with one another but not with others. Some fishers "manage" information within their small group, sharing with them, but not with others, or control information by secrecy to protect their resources (McGoodwin 1990).

The point is that neither image of the fishing community completely matches the reality. Fishing societies are seldom smoothly functioning units, but they do not often match the description in Hardin's "tragedy" either. As a commentary on social relations among fishers, Hardin's "tragedy" is seriously misleading. A fishery management system based on the premise of utility maximizers who exist independent of communities, and who have no kinship relations, social obligations or group memberships, is going to be gravely deficient. It will end up ignoring the norms, networks and trust relationships (also termed social capital), reciprocities, adaptations, values, and local institutions.

Such a misguided fishery management system, among other things, will overlook women's roles and contributions to community economy and civil society. Jentoft (1999) provides examples of the role of women from Scandinavian fishing communities: women's role as "ground crew" in small-scale fisheries in Norway; women's role in providing a range of services that are key to the viability of the fishing household; and the economic contribution of women as a buffer in times of crisis. Women's efforts in a crisis-ridden fishing community in Finnmark, channelled through their local association, Helselaget, kept the community together and maintained spirit and meaning during the period of crisis. Women's contributions were not restricted to the household and their husbands' businesses. They also took on responsibility for the whole community, as community spokespersons addressing the larger society. Such contributions are going to be missed by a fisheries management system that concentrates solely on the fish populations and the fishermen-as-rights-holders. However, a management system that also takes into account the alternate integrated view of fishing communities will note that fishermen could only function within the larger context of the community. Such a management system would also recognize that women and other members of the larger fishing community are also stakeholders in fisheries management. In a real sense, all members of a fishing community can be considered as holders of resource rights.

Jentoft sees overfishing, not always as the result of market forces and poor management, but also, in some cases, the result of community failure. Such community failure can, for example, come about through failure to instil self-restraint, responsibility, and social cohesion among community members. A community that has disintegrated socially will also lose its ability to sanction unacceptable fishing behaviour. The point is perhaps easy to see with reference to the fishing societies reviewed by Wilson and colleagues. If fishery management consisted mainly of rules and practices to regulate how fishing is done, essential social controls on the behaviour of fishers, community failure will clearly lead to the failure of fisheries management. If this is true, then it follows that fisheries management can be improved by designing a management system that consists of more than just rules and regulations. It also consists of community values.

The challenge for the manager is to design mechanisms "that encourage cooperation, build networks, and improve trust within and among local communities" (Jentoft 2000). Management must turn from the consideration of only the fisherman-as-rights-holder, to the consideration of entire communities as stakeholders. Resource rights belong to entire communities, rather than to individual fishermen. Civil institutions of the community, such as community groups or NGOs, are not irrelevant to fisheries management but could be considered to be a part of the larger picture of management. Thus, maintaining the fishing community itself is part of the objective of fisheries management. The task includes building and supporting those processes that improve social cohesion and add to the social capital of fishing communities.

7.5 Institutions and Capacity building

What factors influence the ability of a community to engage in collective action to solve the "tragedy of the commons"? The fishery manager needs to know something about the international experience regarding collective action for fisheries management and building institutions. We deal with institution building as part of capacity building in general. Community-based institutional capacity building is widely recognized as one of the vital components of coastal resources management (Christie and White 1997; Rivera and Newkirk 1997; Pomeroy and Carlos 1997).

The logic of capacity building is simple. Maintaining fishing communities and involving them in the management process depends on the existence of appropriate institutions. But not all fishing communities have the capability to make their own rules or to regulate themselves. Communities in some parts of the world have traditions of social organization and autonomous decision-making for resource management. They may have their own resource use areas and a system for making rules of conduct. However, in the case of other areas and communities, self-organization does not come easily, and it may take effort to organize and build institutions.

In such places, people may be poorly organized above the level of the household. They may not have a history of associations and institutions, and hence little cultural background in collective action. To put it another way, the ability of a community to engage in collective action has to do with the presence of a well functioning civil society. Managers rarely discuss civil society institutions in relation to fisheries management, but the ability of a fishing community to regulate its own affairs is closely linked to whether or not there is a culture of civil society organization in the first place. If civil society traditions are weak, capacity building becomes even more important.

Capacity building is described by the United Nations Development Programme (UNDP) as the sum of efforts needed to nurture, enhance and utilize the skills and capabilities of people and institutions at all levels — nationally, regionally, and internationally — so that they can better progress toward sustainable development. Capacity building is based on a comprehensive view that emphasizes the importance of institutional arrangements, appropriate government policies and legal frameworks, and stakeholder participation. In the context of fisheries, the objective of capacity building is to improve not only the quality of resource management, but also the effectiveness of management performance. Capacity building does not seek to resolve specific resource management problems. Instead, it seeks to develop the capacity within fishing communities, governments, and other organizations (such as NGOs) to resolve their own problems individually or collectively.

7.5.1 Challenges of Capacity Building

A major challenge of capacity building is to reverse the effects of centralized resource management over many generations, which tends to suffocate the ability of fishing communities for self-governance. Top-down resource management over a long period of time can result in the loss of civic institutions and local mechanisms for consensus building, rule making, enforcement, and monitoring. But capacity building does not apply only to fishing communities. All parties need to develop important skills for the cooperative solution of problems: governments, NGOs, professional associations, educational institutions, research groups, and development agencies. Building capacity is a long-term, continuous process. Canada's National Round Table on the Environment and the Economy (NRTEE) identifies four major elements of a strategy to build capacity: improving the knowledge base; developing better policies; enhancing management practices; and reforming institutions (**Table 7.2**). Note that NRTEE's list incorporates not only the above but also elements related to the use of traditional knowledge, user participation, integrated coastal zone management, and partnerships, as discussed in Chapter 2.

How does capacity building work when applied to practical problems of fishery resource management? Two examples are offered, one dealing with an attempt to improve reef fishing practices in Jamaica, and the second dealing with an example of community-based species conservation in Vanuatu.

TABLE 7.2 FOUR MAJOR WAYS TO BUILD CAPACITY.

1. Improve the knowledge base to facilitate better decision-making	Support research by improving data collection, maintenance and analysis, scientific and support practical research, and by incorporating traditional knowledge.
2. Develop better policies and strategies	Reform legislation and policies that hinder the sustainable management of resources and the adoption of integrated management approaches to coastal resources. Raise awareness of sustainable management practices at all management levels.
3. Enhance management practices and techniques	Train professional staff to adapt to new paradigm based on participatory decision-making. Support integrated ocean and coastal zone management in place of the more traditional sectoral approaches. Learn from the lessons of others, and help local institutions to become more self-reliant. Work at all levels to facilitate dispute resolution.
4. Reform institutions	Create partnerships involving user groups, NGOs, the private sector, and government. Strengthen and even create, where necessary, new cooperative arrangements to deal with the impacts of land-based activities on the marine environment.

Source: NRTEE (1998, p. 30).

The north coast of Jamaica supports a very narrow continental shelf and a heavily used reef fishery. Fish populations of the coral reefs of the area are among the most heavily overfished in the Caribbean region. The small-sized wire mesh used to construct traps for reef fish is part of the reason why larger-sized fish are all but absent in the reef area. In 1991, a community-level mesh exchange program was initiated; fishers were given large mesh (1.5 inch = 3.8 cm) chicken wire for trap construction, in exchange for handing in their small mesh (1 and 1.25 inch = 2.5 and 3.2 cm) traps (Sary *et al.* 1997). Monitored over a four-year period, the total number of traps in use in the area and the mean number of trap hauls did not change significantly. The proportion of small-mesh traps did change, from 94 percent (pre-program) to 32 percent (one year later) and to 58 percent (three years later). The results indicate that reef fish populations and hence yields increased somewhat over the four-year period. However, fishers still using small-mesh traps were benefitting the most from the conservation program because small-mesh traps caught more per haul than did the large-mesh traps! The conversion was not sustainable.

The second example deals with the development of village-based marine conservation in the Pacific island nation state of Vanuatu (Johannes 1998b). The valuable invertebrate *Trochus* is readily depleted and has to be harvested with care. When a conservation problem emerged in Vanuatu, the Fishery Department decided not to declare a closed season from the top down, but rather initiate village-based *Trochus* conservation, actively including the villagers in the design and conduct of the program. The villagers were taught the basic principles of *Trochus* management — that stocks of the mollusc should be harvested about once every three years and the fishery closed during the intervening periods. The villages accepted the Department's advice and found that the proposed management worked. The word of their success soon spread to other villages that had not been a part of the initial program. Within a period of four years, many additional villages were managing their *Trochus* stocks and similar conservation methods were being applied to some other species as well.

One of the factors that seem to have contributed to the success of community-based conservation in Vanuatu was the revival of traditional reef and lagoon management rituals and ceremonies. Johannes (1998b) noted that the use of ancient ceremonies and hence traditional marine tenure institutions were crucial for the success of modern conservation in Vanuatu. Properly observed taboos added an emotional or spiritual dimension to the enforcement of conservation measures:

Two village elders told of experiences that have caused them to modify the way in which a fishing taboo is formally declared. When fishing taboos were merely announced without fanfare, observance was unsatisfactory. Now, in these villages, closures are announced with substantial traditional ceremony. Pigs are killed, a feast is held and church leaders are asked to bless the taboo. By thus impressing villagers with the seriousness of the taboos, their observance, according to these leaders, is now much improved (Johannes 1998b).

7.5.2 Analyzing success and failure

The two cases are instructive. The Jamaica case is remarkable in that it shows a free-rider effect: more than two-thirds of the reef fishers initially adopted the large-mesh trap, but this proportion fell to less than a half by the end of the study period. By contrast, conservation compliance was very high among Vanuatu *Trochus* fishers, and furthermore, conservation measures diffused to other villages and to other species as well. Why was there such a difference in the outcomes of the two programs?

Sary *et al.* (1997) do not provide details on how and who designed the mesh exchange program in the Jamaica case, and how much (if any) capacity building accompanied it. There was a fishery cooperative in place in the area, and it was used in the distribution of the equipment (Mahon, field notes). But how were the fishers organized in the community in question? What input did they have into the conservation program? What attempts were made to generate consensus on the action to be taken? Although all fishers would have benefited from increased catches if all had converted to the large-mesh wire, given the large (and increasing) proportion of free-riders, this did not occur.

The community did not buy into the conservation measure strongly enough for collective action to take place. No doubt relevant, Jamaica and the Caribbean in general (because of the colonial background), traditions of cooperation and civil society are weak. As well, there was no provision for alternative income while the stock was rebuilding (R. Mahon, field notes). By contrast, Vanuatu, like many other Pacific islands, has a rich tradition of marine tenure systems, and the practice of closed seasons and species taboos are very much a part of a culture that is being revitalized. Even rituals which help people remember conservation practices are in living memory and relatively easy to rehabilitate.

How do institutions emerge? Sometimes institutions emerge spontaneously, as in the Turkish Mediterranean fishery with the system of rotational fishing sites (Berkes 1986). Several other examples described earlier also fall into this category, including and the Sri Lankan lagoon in **Box 7.4**, US Atlantic coast lobster fisheries (Acheson 1988), Solomon Islands reef and lagoon tenure systems (Baines 1989). Institutional arrangements require an investment of time on the part of the members. As we touched upon earlier, there are transaction costs, or the costs of doing business, to be reconciled against the benefits of collective action. These transaction costs include the costs of getting information, reaching agreements with other group members, and enforcing the agreements that have been reached. If these costs are relatively high, and the benefits of collective action (e.g. better fish yields, less conflict) relatively low, then the conditions are favourable for institutions to emerge. However, if this cost-benefit calculation is not favourable, then institutions will not emerge (Feeny 1998). In such situations, capacity building may help.

We do know that fishery resource management institutions can emerge with development assistance, empowerment and capacity building, as in the St. Lucia mangrove case (Smith and Berkes 1993), the ICLARM projects in Southeast Asia (Katon *et al.* 1997), and in the ICLARM/Ford Foundation projects in Bangladesh (Ahmed *et al.* 1997). Replication of experiences helps enormously, as seen in Oceania, Bangladesh, and East Africa cases. Partnerships among communities, NGOs, and governments to design innovative communication and dissemination techniques, such as the use of videos as done in Tanzania, appear very promising (Moffat et al 1998).

However, it is important to note that capacity and institution building takes years. A Philippines case described by van Mulekom (1999) is instructive in that it provides a timeline of activities for an institutional development process for community-based resource management that shows where capacity building fits in (**Table 7.3**). The case includes the phases of community organizing; capacity development; establishing community-based management; implementing community-based management; and development of non-fishing livelihoods. It shows that the whole process requires on the order of 10 years, consistent with the previously cited experience in St. Lucia and Bangladesh.

Table 7.3 Phases in the development of community-based coastal resource management in Orion, Philippines.

Phase	Activities
1. Community organizing (1989–94)	Advocacy; political action; education; learning-by-doing for project implementation; data gathering; formation of peoples organizations and bay-wide federation of POs
2. Coastal resource management planning and organizational capacity building (1992–99)	Building links with government agencies; training; organizational and institutional capacity building; preparation and presentation of management plan; start monitoring; committees
3. Establishing community-based management system (1994–99)	Dialogue with municipal government; lobbying for municipal ordinances; fund accessing; establishing fisherfolk patrol (for enforcement); formalization of comanagement arrangements
4. Implementing community-based management system (1994-2004)	Document tangible results of plan; artificial reefs; mangrove reforestation; fish sanctuary and fisheries reserve; expand rehabilitation and patrol area; negotiate and later phase-out of trawling
5. Non-fishing livelihood development (1996–??)	Diversifying livelihoods to increase income and reduce fishing pressure; community stores; savings and credit schemes; income-generating projects at the household level.

Source: Adapted from Van Mulekom (1999)

7.6 Conclusions

Making marine commons work and solving the "tragedy of the commons" starts by addressing the two basic characteristics of the commons: how to control access to the resource (the exclusion problem), and how to institute rules among the users (the subtractability problem). Once property rights and resource use rules have been established, both the costs and benefits of any management action will be borne by the same individual or group, thus providing incentive to conserve. The technical literature on common property has established that commons management can succeed or fail, depending on the circumstances of each case, in each of three property-rights regimes, state property, private property, and communal property.

The problem of exclusion may be solved by establishing property rights. Quota harvesting rights (more appropriate for industrial large-scale fisheries) and communal rights such as territorial rights (TURFs) are two of the mechanisms for controlling access. Communal resource tenure systems, such as those in the Tropical Pacific, provide entry points for solving the "tragedy" in coastal waters and pave the way to establish property rights in areas in which resource harvesting had previously operated under non-sustainable, open-access conditions.

Solving the problem of subtractability involves the making and enforcing of resource use rules among the fishers themselves. Laws and regulations are often made by the government. But the experience in a number of areas, such as the coastal fisheries of Japan in which the government has devolved legal rights to village-based Fisheries Cooperative Associations (Ruddle 1989; Weinstein 2000), indicates that the sharing of management power and responsibility has a number of advantages. It can lower costs of management and enforcement and result in more appropriate rules that take advantage of local knowledge. As well, involving the local community in management strengthens community institutions and can lead to more effective fishery management. Capacity building seeks to develop the capability of fishing communities, government agencies and other organizations to solve problems and improve the quality of resource management.

Chapter 8
Comanagement and Community-based Management

8.1 Introduction

Approaches to management and governance of fisheries resources are undergoing a significant transition. There is a shift toward conservation and ecosystem-based management, away from stock- and species-based management. Governance is shifting toward community-based and comanagement approaches, which emphasize fisher participation and decentralization of management authority and responsibility. To illustrate the concept of comanagement, this chapter begins with a case study of a successful comanagement fishery project in the Philippines.

8.2 Conservation project of San Salvador Island, Philippines

The fishery of San Salvador was showing signs of overexploitation in the late 1970s. The fish catch was in decline; illegal fishing using cyanide and explosives was rampant. The fishery was *de facto* open-access, with virtually no law enforcement. From an average reported catch per fishing trip of 20 k in the 1960s, the catch had declined to barely three k in 1988. Many reef fishes, such as groupers, snappers, and damselfish, had become scarce. In 1988, living coral cover had declined to an average 23 percent for the entire island. Though the San Salvador fishers knew that action was needed to protect their livelihood and the resource, the central government of the Philippines was too distant to control the situation and the fishers themselves were too fragmented to embark on any collective action to avert resource degradation.

San Salvador Island, with an area of 380 ha, forms part of Masinloc municipality in the province of Zambales, on the western coast of Luzon, about 250 km north of Metro Manila (**Figure 8.1**). In its population of 1 620 persons, the majority of the 284 households depend on fishing for their livelihood. The island has been home to three generations of residents, the first of whom came from the mainland and were largely farmers who fished part-time using hand lines and nets. During World War II, Japanese troops occupying the island sometimes used explosives to catch fish. After the war, villagers continued with their non-destructive subsistence fishing. Until the late 1960s, resource-use conflicts were rare and the resource remained in good condition. But the early 1970s saw an influx of fishers from the central Philippines who brought illegal fishing methods such as cyanide, fine mesh nets, and explosives. The new fishers also integrated the village economy into the international market for aquarium fish.

In the late 1980s, when resource overexploitation, degradation and use conflicts reached a crisis point in San Salvador, residents went in search of solutions to their problems. External change agents were instrumental in initiating new resource management measures. A Peace Corps volunteer who arrived in San Salvador in 1987 conceptualized the Marine Conservation Project for San Salvador (MCPSS), a community-based coastal resource management (CBCRM) project for coral reef rehabilitation. In 1989, a local non-governmental organization led a project to establish a marine sanctuary. The project featured biological (sanctuary and reserve) and

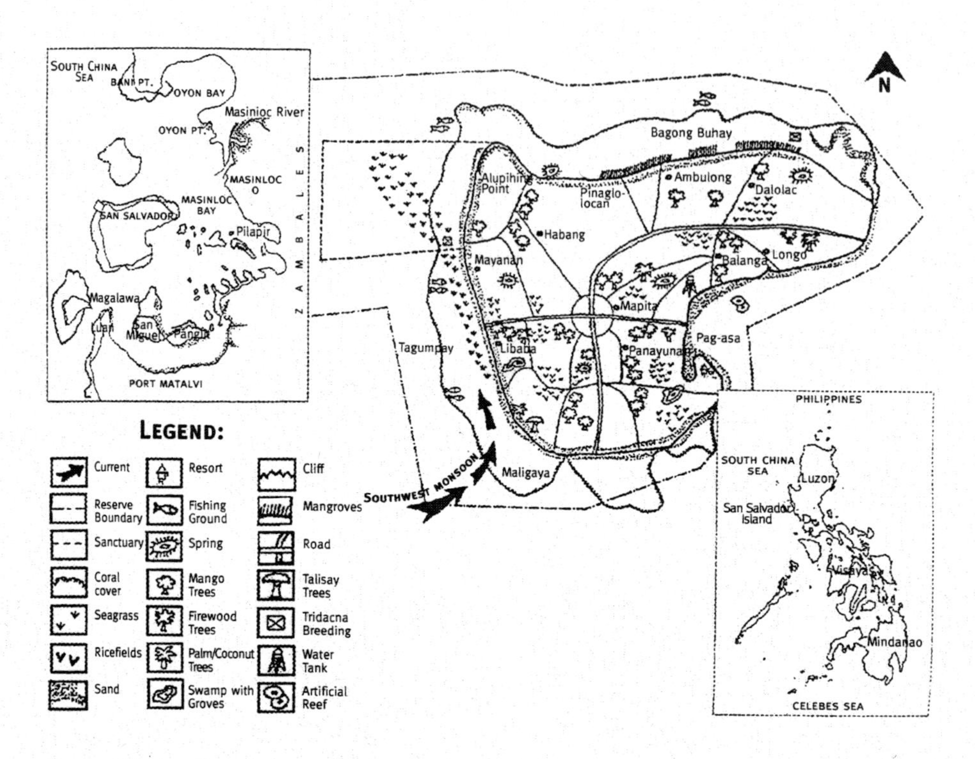

Source: Haribon Foundation

Figure 8.1 Map of San Salvador Island.

governance interventions (management plan, community organizing, income-generation, rules and regulations, education, and training). That same year, the core group members made an exchange visit to a successful marine sanctuary in the central Philippines. The visit increased motivation and support for the idea of a sanctuary and reserve, and resulted in the drafting of a local ordinance to ban fishing within the sanctuary and allow only non-destructive fishing methods in the reserve. In July 1989, the Masinloc Municipal Council passed an ordinance for the marine sanctuary and reserve. Core group members became increasingly active in monitoring illegal fishing activities and guarding the sanctuary. Other resource users participated in village consultations, endorsed local ordinances, adhered to the rules, and adopted non-destructive fishing methods.

While the MCPSS was not conceived as a comanagement project, the increased role and participation of the government brought about a resource management partnership between government and fishers. The Masinloc municipal government, which has political jurisdiction over San Salvador, was drawn into the picture in a number of ways: 1) passed enabling legislation that provided a legal basis for the sanctuary and for apprehending rule violators; 2) mediated conflicts between local and outside resource users; 3) provided boat and equipment for patrolling coastal waters; 4) created a government patrol team to enforce laws; and 5) provide a political environment that allowed for the pursuit of community-based initiatives. Thus, comanagement can be considered to date back to mid-1989, prompted by the political dynamics in San Salvador and the village fishers' lack of resources to run enforcement activities. The main government partner in this comanagement arrangement was the municipal government; the support of the national government through the national fisheries agency was not as visible.

In 1991, policy and legal support for comanagement was strengthened in the Philippines through passage of the *Local Government Code*, which gave the municipal government jurisdiction over nearshore waters. Following the turnover of the San Salvador Island project from the NGO to the village-based fisher organization in 1993, comanagement became increasingly visible. Fishers and government shared responsibility for law enforcement, and the government provided funds for local enforcement operations.

San Salvador Island comanagement was a win-win solution. For the resource users, comanagement was prompted by their dependence on fishery resources for livelihood, recognition of resource management problems, and legitimacy and enforceability of rules. The government was motivated by its concern for improved living conditions for the fishers and their families and for sustainable resource management. The path to comanagement was not trouble-free, however. Fishers using destructive methods, and those displaced from the sanctuary, became alienated and resentful. Over time, however, tangible benefits in the form of higher fish catch from San Salvador's fishing grounds helped to encourage rule compliance and non-destructive fishing practices.

A large number of factors contributed to the success of the project. These included the resource stakeholders' participation and sense of ownership in project planning and implementation, clarity of objectives, supportive leadership, the partnership between fishers and government, specification and legitimacy of user rights and enforcement, capability building, and tangible benefits such as redefined resource access, a shift to non-destructive fishing methods, improved enforcement, and observable biological, economic and social changes. Tangible benefits were clearly important. Fish catches went from barely three k in 1988 to six to 10 k in 1998, accompanied by an improvement in the diversity of fish species. The extent of living coral reef cover increased from 23 percent to 57 percent for the whole island. Fishers perceived gains in equity, knowledge, household income, empowerment, and conflict reduction. The San Salvador comanagement project gave the village residents a reason for optimism, a motivation for collective action, and pride in their resource management achievements.

8.3 Why comanagement?

The case of San Salvador Island is but one example of successful comanagement. But since the comanagement approach is relatively new, we must ask questions about its broader applicability. Why do we need comanagement? Why do we need a shift in fisheries management strategies?

8.3.1 The search for better management approaches

The last 50 years have seen philosophies shift in the fisheries development and management process. The period after World War II was one of reconstruction of the world's fishing fleets. The 1960s saw expansion, with the opening up of new fishing grounds and development of new technologies and long-range fleets. In the 1970s, fishing continued to grow but overfishing was increasingly noticed. In the early 1980s, the United Nations Conference on the Law of the Sea brought about coastal countries' expansion of exclusive economic zones (EEZ) to 200 miles. Expanded EEZs caused a redistribution of access to ocean resources and use-rights, and brought a great deal of ocean space under single-nation resource management. Throughout the 1980s and early 1990s, global concern grew about resource overexploitation, environmental degradation, and threats to biodiversity, and a call went out for sustainable development. The 1990s brought several international initiatives, including the UN Conference on Environment and Development, the *International Convention of Biological Diversity*, the *International Plan of Action for the Management of Fishing Capacity*, and the *Code of Conduct for Responsible Fisheries*. These challenged countries to encompass sustainable utilization of fisheries resources (FAO 1996a, 1997; Garcia 1994; Prado 1997). A central element of these initiatives is the increased participation of resource users, changed from merely being consulted and receiving top-down information to participating in decision-making and interactive management.

The changing philosophies of the fisheries development process are reflected in changing approaches to fisheries resource management. Before colonialism, various kinds of traditional and customary fisheries management regimes were in place in most countries. Some coastal societies, especially in the Asia–Pacific region, had sophisticated traditions, customs, and sea tenure systems consistent with conservation (Ruddle *et al.* 1992). Other areas were simply managed by default, since small populations and simple technology did not overexploit marine resources (Johannes 1982).

During the colonial period, governance of coastal and marine resources was transferred from communities to local and national government bodies (Pomeroy 1995). In most colonies, centralized management agencies were established to control the level of exploitation, modernize fishing methods, and ensure exports back to the colonizing country. The centralized approaches to management that began centuries ago in some countries continued under the neo-colonial regimes of newly independent nations as they consolidated power. By appropriating control over fisheries management, national governments often underestimated coastal communities' capacities, often learned through long and difficult experience, to manage local fisheries to meet their needs. In many instances, the national government overestimated its ability to manage these same resources. When community-level institutional arrangements for coastal fisheries management are undermined, the usual common-property resource management regimes have been replaced, in many cases, not by science-based government management but by open-access regimes.

Among Western-trained fisheries managers, resources management fisheries has been based on the conventional wisdom that it is possible to manage fisheries successfully by keeping three facts in mind: 1) when left to their own devices, fishers will overexploit stocks; 2) those stocks are extremely unpredictable; and 3) to avoid disaster, managers must have effective hegemony over them (Berkes 1994b). The conventional or centralized management approach has been dominated by the assumption that every fishery is characterized by intense competition, which will eventually lead to overexploitation and the eventual dissipation of resource rents: the so-called "tragedy of the commons" (see Chapter 7). It also relies almost exclusively on scientific information and methods, as opposed to traditional and customary knowledge and management systems. This has led managers in the direction of tighter government controls over fisheries. Over time, these controls have become complicated, costly, and unworkable.

Due to the recent failure of so many fisheries, the conventional management approach has been widely called part of the problem rather than of the solution of resource overexploitation. On the biological side, the traditional approach fails to take into account the ecological complexities, especially in the case of tropical fisheries. On the human side, bureaucrats and professionals have replaced resource users as resource managers. The fishers do little to monitor and police themselves. The centralized management approach, which makes little or no use of fishers' capacity to manage themselves and does little effective consultation of the resource users, is often not suited for developing countries with limited financial means and expertise to manage fisheries resources in widely dispersed fishing grounds.

8.3.2 New Directions in Fisheries Management

In the last decade, following concerns about conventional management as well as fishery overexploitation and environmental degradation, the objectives, approaches, and policies of fishery management systems have begun to change. The objectives have shifted from maximizing annual catches and employment to sustaining stocks and ecosystems, and from maximizing short-term interests to addressing both short- and long-term interests (see Chapter 1). There is a shift away from conventional production and stock- and species-based management toward conservation and ecosystem-based management. Policies have shifted from open and free access, sectoral fishery policy, command-and-control instruments (the use of various harvest control regulations), and top-down and risk-prone approaches to limited entry, user rights and user fees, coastal zone intersectoral policy, command-and-control and macro-economic instruments, and participatory and precautionary approaches (Garcia 1994). Governance of fisheries is shifting toward the use of market regulation on the one hand, and community-based management and comanagement on the other. It is increasingly recognized that resources can be better managed when fishers and other stakeholders are more involved in management of the resources and when use rights are allocated — either individually or collectively — to control access. Devolution of management authority and responsibility is bringing about shifts in local power elites and structures. These new approaches will require changes in the administrative levels of management, as well as new laws and policies to support the new management arrangements.

8.3.3 Property rights

These alternative approaches are meant to deal with the "perverse economic incentive system" that arises largely from the ill-defined resource property rights that characterize capture fisheries resources (Munro *et al.*1998). From an economic perspective, the causes of overfishing are generally found in the absence of property rights or other institutions that might otherwise provide exclusive control over harvesting and, as a result, an incentive to conserve. These alternative approaches range from community-based management and comanagement, meant to address the lack of participation and conflicts that were the legacy of centralized management, to market regulation and rights-based management, which are meant to reduce excess competition and investment and incite the fisher to greater economic efficiency. In fisheries, the prime example of market regulation is the individual transferable quota (ITQ) system. ITQs carve up the total allowable catch into private (property) rights that can be freely traded in a market. The premise behind the market regulation approach is that the profit orientation and cost-consciousness of individual fishers will result in the most efficient method of catching fish for the society as a whole.

8.3.4 Refocusing management: Participation

It is interesting to note that fisheries managers had, until recently, been tightening government controls, whereas those in other fields of management had been moving in the direction of devolution, deregulation, decentralization, and comanagement (Berkes 1994b). This slowness in moving in new directions in fisheries may be due in part to the complexity of the natural and human ecosystems in marine and coastal environments.

Fisheries managers are increasingly recognizing that the underlying causes of fisheries resource overexploitation and environmental degradation are often of social, economic, institutional and/or political origins. The primary concerns of fisheries management, therefore, should be the relationship of fisheries resources to human welfare and the conservation of the resources for future generations to use. That is, fisheries management should focus on people, not fish, per se. Policy interventions, if they are to bring about lasting solutions, must address these concerns.

National governments, for the most part, have failed to develop an adequate substitute for traditional resource management systems. The promotion of nationalization and privatization as routine policy solutions has not solved the problem of resource overexploitation and, in many instances, has deprived large portions of the population of their livelihood (Bromley and Cernea 1989). Under these conditions, the devolution of fisheries management and allocation decisions to the local fisher and community level may be more effective than the management efforts that distant, understaffed, and under-funded national government fisheries agencies can provide.

Fishers, the real day-to-day managers of the resource, must be equal and active participants in fishery management. An open dialogue must be maintained between all stakeholders. Property rights to the resource must be assigned directly to its stakeholders — the coastal communities and resource users. The "community" must be reinvigorated through multi-sector, integrated resource management and social, community and economic development (for example, improved social services and infrastructure, new employment and livelihood opportunities) (see **Box 8.1**). A new management philosophy is warranted, one in which the fisher can once again become an active member of the resource management team, balancing rights and responsibilities, and working cooperatively, rather than antagonistically, with the government. Such cooperative or joint management — comanagement— is a logical extension of the evolution of fisheries management in recent decades.

Box 8.1 Management and Development.

Because the terms "management" and "development" can mean different things to different people, we will define each.

Management refers to activities undertaken to protect, conserve and rehabilitate a resource, including policy action such as regulations and material interventions such as artificial reefs.

Development refers to activities that develop a resource and its users and stakeholders. Fisheries development increases production from the resource. Economic development improves employment and livelihood opportunities through economic growth. Community development improves community services (education, health) and infrastructure (roads, water, communication). Social development improves people's capacity to participate in governance and find local solutions to problems and opportunities.

8.3.5 So, why comanagement?

In summary, why comanagement? Fisheries management should be done on a small ecological scale to be effective. Attention must be paid to local ecological factors, such as habitats and local populations, central to the health of the whole ecosystem, both biological and human. Fisheries management will need to be designed to fit this smaller scale, focusing on local-level management, decentralization of management authority and responsibility, and use of fishers' knowledge. Fishers can no longer depend on government to solve their problems, whether community or fisheries-related. They will have to take more responsibility for management and be accountable for their decisions. This means bearing the costs of benefiting from those decisions.

There are several other reasons for the interest in and push for comanagement worldwide in addition to those mentioned above. The crisis in fisheries and coastal communities is pressuring national governments to look for alternative management strategies. Many governments view comanagement as a way to deal with this crisis

Box 8.2 The role of NGOs in comanagement.

Nongovernmental organizations (NGOs) play an increasingly central role in coastal resource conservation and management programs. This critical role is being assigned by international donors and development agencies that designate the implementation of community projects to NGOs and people's organizations. The role is also being assigned by governments, such as in the Philippines, where the *Local Government Code* of 1991 gives a legal mandate to NGOs to participate in the local development council and, under the *Fisheries Code* of 1998, to participate in local aquatic resources management councils. This recognition is the result of the effectiveness and dynamism of NGOs to organize and empower people.

While NGOs have many roles, depending upon their motivation and perspective, one of their central roles in comanagement is facilitation. They work with the people in a community to set up the social infrastructure necessary for comanagement. The NGO is a partner and change agent, providing information and independent advice, ideas and expertise, education and training, and guidance for joint problem solving and decision-making, thus enhancing the people's ability to manage their own lives and resources.

Problems can arise when people become too dependent upon the NGO or when the NGO directly interferes in the process, rather than guiding it or serving as a catalyst. Problems can also occur when the NGO's ideological views on development are not acceptable to the community or government. The role of NGOs in coastal resource conservation and management programs differs country by country. In Brazil and Turkey, for example, NGOs do not play a major role in these programs.

Source: Anonuevo 1994

by passing responsibility to the resource users, thus lessening government's burden of cost and responsibility for resource management (A problem with this is that the resource users are usually given a severally degraded and overexploited resource to manage.) International donors and development agencies constitute another force driving comanagement They like the concept of participation and so encourage governments and NGOs to implement it In many countries, NGOs are at the forefront of implementing comanagement in coastal communities (**Box 8.2**) And in some areas, certain groups or powerful individuals see comanagement as a way to get increased access and control over resources. Once the process is initiated they try to manipulate the system in their favour

8.4 What is fisheries comanagement?

In the wake of the historical record of often ineffective centralized fisheries manage ment is the need to change the structure of governance. What is needed now is a more dynamic partnership using the capacities and interests of the local fishers and com munity, complemented by the state's ability to provide enabling legislation, enforcement and other assistance This approach to fisheries management will require a shift away from a centralized, top down form of management to a new strategy in which fisheries managers and the fishers jointly manage the fisheries: 'comanagement" (Jentoft 1989; Pinkerton 1989a; Berkes *et al* 1991; Berkes 1994b) Comanagement includes a sharing of governance structures between stakeholders in the resource and institutions of local collective governance of common property (**Figure 8 2**)

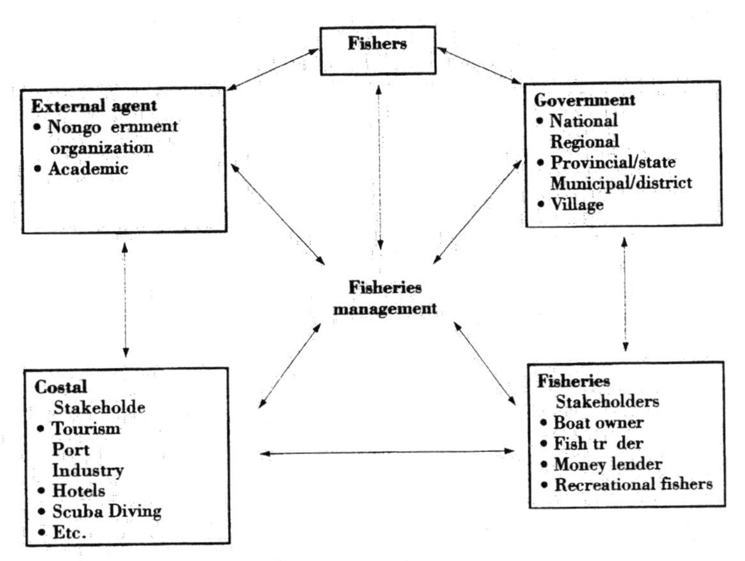

Figure 8 2 Fisheries comanagement is a partnership.

8.4.1 Definition

Fisheries comanagement can be defined as a partnership in which government, the community of local resource users (fishers), external agents (non-governmental organizations, academic, and research institutions), and other fisheries and coastal resource stakeholders (boat owners, fish traders, money lenders, tourism establishments, etc.) share the responsibility and authority for making decisions about the management of a fishery. This partnership can be seen on a continuum between purely government-based management and community-based management (see **Figure 2.3** in Chapter 2). Through consultation, the partners develop an agreement that specifies their roles, responsibilities, and rights in management. Resource users assume an active and constructive role in management and stewardship of the resource. Comanagement covers various partnership arrangements and degrees of power sharing and integration of local (informal, traditional, customary) and centralized government management systems. There is a hierarchy of comanagement arrangements (**Box 8.3**), from those in which government consults fishers before introducing regulations, to those in which the fishers, with advice from the government, design, implement, and enforce laws and regulations (Sen and Raakjaer-Nielsen 1996). It is generally acknowledged that not all responsibility and authority should be vested in the local level. The amount

Box 8.3 Levels of comanagement.

There is a continuum of power sharing of authority and responsibility between government and community, ranging from minimal exchange of information to community control through delegation of management authority. The degree of comanagement increases along this range from informative to community control on the continuum.

Community control	Power delegated to community to make decisions and inform government of these decisions
Partnership	Partnership of equals with joint decision-making
Advisory	Users advise government of decisions to be taken and government endorses these desisions
Communicative	Two-way information exchange; local concerns are represented in management plans
Cooperative	Community has input into management
Consultative	Mechanism exists for government to consult with fishers; government makes all decisions
Informative	Community is informed about decisions that government has already made

Source: Berkes 1994a; Sen and Raakjaer-Nielsen 1996

of responsibility and/or authority that the state level and the various local levels have will differ, depending upon country and site-specific conditions. Determining what kind and how much responsibility and/or authority to allocate to the local levels is a political decision. Furthermore, the government will always hold the balance of power in comanagement.

8.4.2 Comanagement as a process

Given the different conditions, processes, needs, and demands within the fisheries sector, there is no simple management solution appropriate for every community, region, or nation. There is no blueprint for comanagement but rather a variety of arrangements from which to choose to suit a specific context. Comanagement should be viewed not as a single strategy to solve all the problems of fisheries management, but rather as a process of resource management, maturing and adjusting to changing conditions over time, and involving aspects of democratization, social empowerment, power sharing, and decentralization. Comanagement is adaptive; through a learning process, information is shared among stakeholders, leading to continuous modifications and improvements in management. Comanagement is not a regulatory technique. It is a participatory and flexible management strategy that provides and maintains a forum or structure for action on participation, rule making, conflict management, power sharing, leadership, dialogue, decision-making, negotiation, knowledge generation and sharing, learning, and development among resource users, stakeholders and government. Comanagement is a consensus-driven process of recognizing different values, needs, concerns, and interests involved in managing a resource. Partnerships are pursued, strengthened, and redefined at different times in the comanagement process, depending on the existing policy and legal environment, the political support of government for community-based actions and initiatives, and community organizations' capacities to become partners of government. The comanagement process may include formal and/ or informal organizations of resource users and stakeholders. The establishment and successful operation of fisheries comanagement can be complex, costly, multiyear, and sometimes confusing (Rivera 1997; Pomeroy *et al.* 1999). A review of over 100 community-based coastal resource management (CBCRM) projects in the Philippines revealed that the average duration for projects was 4.2 years (Pomeroy and Carlos 1997). Discussions with NGOs implementing the CBCRM projects found that it takes three to five years to just organize and initiate activities and interventions.

8.4.3 A partnership

Comanagement involves various degrees of delegation of management responsibility and authority between the local level (resource user, stakeholder, community) and the state level (national, provincial, municipal, village government) (**Figure 2.3**). Comanagement is a middle course between the state's concerns about efficiency and equity and local concerns for self-governance, self-regulation, and active participation. Comanagement involves a formal or informal agreement with government to share power and to share the right to manage. Comanagement can serve as a mechanism not only for fisheries management but also for community economic and social development as it promotes fisher and community participation in solving problems and addressing needs. In some cases, comanagement may be simply a formal recognition of a fisheries management system that already exists; some informal and customary community-based management strategies are already in place, side-by-side with formal state-level management strategies.

8.4.4 Stakeholder involvement

Other than fishers, stakeholders that derive economic benefit from the resource (for example, boat owners, fish traders, business suppliers, police, politicians, consumers) should also be considered in the comanagement arrangements (**Box 8.4**). These stakeholders often hold considerable political influence in the resource management regime. A proper balance of representation among the stakeholders will prove crucial to the success of comanagement. A central question, however, is which stakeholder groups should be represented and how those representatives should be chosen. While it is useful to have representation of all stakeholders, a line must be drawn or the process will break down from the representation of too many interests. This question can be partially answered by determining the spatial scale at which comanagement should operate. The best opportunity for comanagement occurs at the local or "community" scale (**Box 8.5**).

Box 8.4 Stakeholder analysis.

Stakeholders are individuals, groups, or organizations of people who are interested, involved or affected (positively and negatively) by a management or development project. They are motivated to take action on the basis of their interests or values. This may originate from geographical proximity, historical association, dependence for livelihood, institutional mandate, economic interest, or a variety of other concerns. Stakeholders are important because they can support and sustain a particular resource; they can be potential partners or threats in the management and development of coastal resources. Different stakeholders generally possess different interests, different ways of perceiving problems and opportunities about coastal resources, and different approaches to management. They should all be equitably represented when a management system is being developing.

The process of identifying stakeholders and figuring out their respective importance regarding decisions on the resource at stake is called stakeholder analysis. This method provides insights about the characteristics of individuals and/or groups and their respective relationship to a resource or project. It also examines the stakeholders' interests in a resource or project and the impact of the activity on the stakeholder. Such analysis is usually conducted in a participatory way.

Source: IIRR 1998; Langill 1999; Borrini-Feyerabend 1997

Box 8.5 Community.

The term "community" can have several meanings. Community can be defined geographically by political or resource boundaries or socially as a community of individuals with common interests. For example, the geographical community is usually a village political unit (the lowest governmental administrative unit); a social community may be a group of fishers using the same gear type or a fisher organization. A community is not necessarily a village, and a village is not necessarily a community. Care should be also taken not to assume that a community is a homogeneous unit, as there will often be different interests in a community, based on gender, class, ethnic, and economic variations. Recently, the term "virtual community" has been applied to non-geographically based communities of fishers. Similar to the "social community," this is a group of fishers who, while they do not live in a single geographical community, use similar gear or target a the same fish species.

Source: NRC 1999

8.4.5 Equity

Comanagement seeks equity in fisheries management. It seeks to empower the weak or less privileged groups in a community to allow them to freely participate in and collaborate on management. Comanagement strives for more active fisher participation in the planning and implementation of fisheries management. Responsibility means fishers have a share in the decision-making process and bear the costs of getting the benefits of those decisions. The theme of comanagement is that self-involvement in the management of the resource will lead to a stronger commitment to comply with the management strategy and sustainable resource use. The mutuality of interests and the sharing of responsibility among and between the partners will help to narrow the distance between resource managers and resource users, bringing about closer compatibility of the objectives of management.

Comanagement provides for collective governance of common property resources (see Chapter 7). As Jentoft (1989) put it, "How then is comanagement to be distinguished from other common property management systems, such as government regulation or community-initiated regulation?" The answer is that comanagement is a governance arrangement between pure state property and pure communal property regimes. The four property rights regimes (state property, communal property, private property, and open access) are ideal, analytical types; they do not exist in the real world. Rather, resources tend to be held in overlapping combinations of these four regimes (see Chapter 7). Strictly speaking, pure communal property systems are always embedded in state property systems and state law, deriving their strength from them. It should be noted that while state law can enforce or strengthen communal property, it might not always do so. The level of help from the state will depend on its willingness to support communal property systems.

8.4.6 CBRM AND COMANAGEMENT

A central element of comanagement is community-based resource management (CBRM), the advantages of which have been well documented in various parts of the world. Community-based fisheries management (CBFM) tends to be more difficult than CBRM for some other resource types because of the complexity of fisheries and aquatic resource systems, the social and cultural structures of fishing communities, the migratory nature of the resource, and the independent nature of fishers.

There is some debate over the similarities and differences between comanagement and CBRM. While there are many similarities between the concepts, the focuses of each strategy differ. These differences centre on the level and timing of government participation in the process. CBRM is people-centred and community-focused, while comanagement focuses on these issues plus on a partnership arrangement between government and the local community and resource users. The process of resource management is organized differently too, comanagement having a broader scope and scale than CBRM. The government may play a minor role in CBRM; comanagement, on the other hand, by definition includes a major and active government role.

Government serves a number of important functions that include provision of supporting policies and legislation, such as decentralization of management power and authority, the fostering of participation and dialogue, legitimization of community rights, initiatives and interventions, enforcement, addressing problems beyond the scope of the community, coordination at various levels, and financial and technical assistance. Government provides legitimacy and accountability to CBRM through comanagement; it must establish commensurate rights and conditions and devolve some of its own powers for both comanagement and CBRM to be effective resource management strategies. Only government can legally establish and defend user rights and security of tenure at the community level.

Comanagement often addresses issues beyond the community level, at regional and national levels, and of multiple stakeholders, and allows these issues, as they affect the community, to be brought more effectively into the domain of the community. CBRM practitioners sometimes view government as an external player to be brought into the project only at a late stage, or as needed. This can lead to misunderstandings and lack of full support from government. Comanagement strategies, on the other hand, involve government agencies and resource managers early and equally, along with the community and stakeholders, developing trust between the participants.

8.4.7 CATEGORIES OF COMANAGEMENT

Based on the above discussion on comanagement and CBRM and on the comanagement literature, it is possible to develop two categories of comanagement: 1) community-centred comanagement and 2) stakeholder-centred comanagement.

When CBRM is considered an integral part of comanagement, it can be called community-centred comanagement. Community-based comanagement includes the characteristics of both CBRM and comanagement; that is, it is people-centred, community-oriented, resource-based, and partnership-based. Thus, community-centred comanagement has the community as its focus yet recognizes that to sustain such action, a horizontal and vertical link is necessary. Successful comanagement and meaningful partnerships can only occur when the community is empowered and organized. This category of comanagement is more complex, costly, and time-consuming to implement than pure CBRM due to the need to develop partnerships early in the process and to maintain them over time. Examples of community-centred comanagement can be seen all over the world, including in Asia (Pomeroy and Pido 1995; Pomeroy 1995), Africa (Normann *et al.* 1998), and the Caribbean (Brown and Pomeroy 1999). Community-centred comanagement seems to be found most often in developing countries due to their need for overall community and economic development and social empowerment, not just resource management.

One variation of community-centred comanagement is traditional or customary comanagement. Such systems are or were used to manage coastal fisheries in various countries around the world. Existing examples in Asia and the Pacific have been documented over a wide discontinuous geographical range (Ruddle 1994). Many of these systems play a valuable role in fisheries management and will be useful into the future, locally and nationally. Ruddle (1994) points out:

> *In many locations, legal issues are among the principal constraints on the viability or future usefulness of traditional marine management systems. Thus, if the contemporary usefulness of such systems has been formally recognized by government, they will require support by appropriate amendments to national laws, and lower-order governments, such as provinces/states, with the explicit and easily understood recognition of customary law and community-based management rights as local corporate entities, accompanied by procedures for establishing the recognition of these rights.*

Traditional or customary comanagement is a formal government recognition of the informal system as done, for example, in Vanuatu and Fiji. Comanagement can serve as a mechanism to legally recognize and protect these traditional and customary systems and to specify authority and responsibility between the community and government. It also involves a definition of shared powers and authority.

Stakeholder-centred comanagement seems to be more common in developed countries, where the emphasis is to get the users participating in the resource management process. It can best be characterized as government–industry partnership that involves user groups in the making of resource management decisions. This category of comanagement focuses on having fishers and other stakeholders represented through various organizational arrangements in management. Unlike in community-centred comanagement, little or no attention is given to community development and social empowerment of fishers. Examples of stakeholder-centred comanagement can be seen in many developed countries in Europe and North America, including Norway's Lofoten Islands cod fishery and the Regional Fisheries Management Councils of the United States (Jentoft and McCay 1995; Nielsen and Vedsmand 1995; Hanna 1996; McCay and Jentoft 1996).

8.4.8 Advantages and limitations

The potential advantages of comanagement include a more open, transparent, and autonomous management process. Comanagement can be more economical than centralized systems, requiring less to be spent on administration and enforcement. In the self-management involved, fishers take responsibility for a number of managerial functions, allowing the community to develop a flexible and creative management strategy that meets its particular needs and conditions and that it sees as legitimate. Comanagement is adaptive, allowing for adjustments in activities in line with the results obtained and lessons learned. Community members understand their problems, needs and opportunities better than outsiders do, so fishing communities are able to devise and administer regulatory instruments that are more appropriate to local conditions than are externally imposed regulations. Comanagement can make maximum use of indigenous knowledge and expertise to provide information on the resource base and to complement scientific information for management.

Management is accountable to local areas, not just to larger regions, and groups of co-managers share joint accountability. By giving the fishers a sense of ownership over the resource, comanagement provides a powerful incentive for them to view the resource as a long-term asset rather than to discount its future returns. Fishers are given an incentive to respect and support the rules because they complement local cultural values, are self-imposed, and are seen as individually and socially beneficial.

Since the community is involved in the formulation and implementation of management measures, a higher degree of acceptability and compliance can be expected: community members can enforce standards of behaviour more effectively than bureaucracies can. Its strategies can minimize social conflict and maintain or improve social cohesion in the community. Empowerment, through information, training, and education, allows the fishers to share power with political and economic elites and government.

Comanagement is required for every fishery (even if that means at the lowest level of consultation) but may not be suitable for every fishing community. Many communities may not be willing or able to take on the responsibility of comanagement. A long history of dependency on government may take years to reverse. Leadership and appropriate local institutions, such as fisher organizations, may not exist within the community to initiate or sustain comanagement efforts. For many individuals and communities, the incentive(s) — economic, social and/or political — to engage in comanagement may not be present. The risk involved in changing fisheries management strategies may be too high for some communities and fishers. The costs for individuals to participate in comanagement strategies (time, money) may outweigh the expected benefits. Sufficient political will may not exist among the local resource stakeholders or in the government, and there is no guarantee that a community will organize itself into an effective governing institution. Actions by user groups outside the immediate community may undermine or destroy the management activities undertaken by the community. Particular local resource characteristics, such as fish migratory patterns, may make it impossible for the community to manage the resource. In addition, the need to develop a consensus from a wide range of interests may lengthen the decision-making process and result in weaker, compromised measures. Comanagement may result in shifts in power bases that are not in the best interests of all partners, and may even result in increased bureaucracy and regulation.

8.4.9 Collective action

Fishers' ability to organize for collective action has a number of prerequisites involving local institutions, defined here as the set of rules actually used (rules-in-use) by a group of individuals to organize their activities (Ostrom 1990; North 1990). Not all groups of fishers have appropriate local institutions; in such cases, any comanagement initiative will necessarily start with institution building. But institution building is a long-term and costly process. Community organizing can take from three to five years to put a self-sufficient organization in place, according to cases in the Philippines (Carlos and Pomeroy 1996). The development of institutions for self-governance takes in the order of 10 years, according to some cases in St. Lucia, West Indies (Smith and Berkes 1993) and Bangladesh (Ahmed *et al.* 1997). The coastal fishery of Alanya, on the Mediterranean coast of Turkey, developed locally designed rules for resource allocation and conflict reduction, which made use of rotating turns at fishing sites. This development took 10 to 15 years, without government support or any other institution-building intervention (Berkes 1986).

8.4.10 The comanagement agreement

At the heart of comanagement is a formal, negotiated agreement between the partners. A comanagement agreement is essentially a management plan that specifies the objectives, partners to the agreement, and rights and responsibilities of the partners. The agreement usually identifies:

- A territory (or set of resources) and its boundaries;
- The range of functions and sustainable uses it can provide;
- The recognized stakeholders;
- The functions, rights, and responsibilities of each stakeholder;
- An agreed set of management priorities and a management plan;
- Procedures for dealing with conflicts and negotiating collective decisions about all of the above;

- Procedures for enforcing such decisions;
- Rules for monitoring, evaluating, and reviewing the agreement and management plan (Borrini-Feyerabend 1997).

The agreement should be dynamic and adaptive. As the comanagement process matures over time, the agreement should be adjusted to reflect partners' changing roles, rights, and responsibilities. A management body with joint authority usually represents the partners. Each partner has clearly defined functions in decision-making. Though the agreement is an essential element of comanagement it can be frightening to the partners, especially the government, because it holds them accountable to meet the specified conditions.

8.4.11 The role of government

The delegation of significant authority to manage the fisheries may be one of the most difficult tasks in establishing comanagement systems. While governments may be willing to call for more community involvement, they must also establish commensurate rights and authorities and devolve some of their own powers (**Box 8.6**). Fisheries administrators may be reluctant to relinquish their authority or parts of it, fearing infringement by local fishers and their representatives upon what they consider their professional and scientific turf. In all cases of comanagement, though, while responsibility is shared, the government holds the ultimate authority.

It should be noted that government does not always undertake comanagement with selfness intentions. It may decentralize management authority and responsibility as a result of poor central government fisheries management efforts and/or a resource crisis.

For example, a community facing an overfished and degraded marine environment complains about the lack of central government management; government, unable or unwilling to address the issue, passes the problem to the community through a comanagement arrangement. Such a community is often left without the capacity or funds to manage the resource.

Government agencies may also engage in comanagement to gain power and the attention of higher government officials by showing how through "comanagement" they are "empowering" fishers to participate in management. A similar dynamic occurs in the case of donor-driven comanagement. Even though local "organizations" may exist only on paper and have no credibility with the fishers, the government agency can claim results — the existence of local organizations — even though no real, active community partnership exists. Moffat *et al.* (1998) commented that some East African cases show this pattern. When donor funding ends, partnerships collapse.

One fundamental debate in comanagement is whether resource users can be entrusted to manage their resources (Berkes 1994b). Unless governments and decision-makers who implement government policies can be convinced of users' desire and ability to manage themselves, not much progress can be made in comanagement. As already stated above, government resource managers are often reluctant to share

Box 8.6 Decentralization and comanagement.

Decentralization is the delegation of power, authority, and responsibility from the central or national government to lower levels, or smaller units, of government, such as states or provinces, or to local-level institutions, such as community organizations. Comanagement requires the central government to be clearly committed to sharing power with local government and organizations.

Decentralization can be operationalized in four ways:

* Deconcentration: the transfer of authority and responsibility from the national government departments and agencies to regional, district, and field offices of the national government.

* Delegation: the passing of some authority and decision-making powers to local officials. The central government retains the right to overturn local decisions and can, at any time, take these powers back.

* Devolution: the transfer of power and responsibility for the performance of specified functions from the national to the local governments without reference back to central government. The nature of transfer is political (by legislation), in contrast to deconcentration's administrative transfer; the approach is territorial or geographical rather than sectoral.

* Privatization: the transfer of responsibility for certain governmental functions to NGOs, voluntary organizations, community associations, and private enterprises.

Source: Pomeroy and Berkes 1997

authority. However, it would be a mistake to interpret this solely as a self-serving motive to hang onto political power. Many managers have well-considered reasons to be skeptical about local-level management. To convince them that this can work, part of the responsibility falls on the resource users themselves. The ability to gain self-management, in turn, partially depends on the local community's ability to control the resource in question.

Managers' reasons for skepticism include the fishers' lack of appropriate knowledge and know-how and the question of the fishers' ability to organize themselves to manage for long-term sustainability. Each of these points opens up a debate. Even in countries with high standards of education, fishers tend to have lower levels of formal education than the general population. But the relevant knowledge about a fishery is not the same as formal education. It is well known that the knowledge held by fishers in many areas of the world, especially in traditional societies in which such knowledge accumulates by cultural transmission, may be extremely detailed and relevant for resource management (Johannes 1981; Freeman *et al.* 1991; Berkes *et al.* 1995). Indeed, it is the complementarity between such local knowledge and scientific knowledge that makes comanagement stronger than either community-based management or government management.

A key question in comanagement is what management functions are best handled at the local or communal level and which at the national government level. Often, these functions are set too narrowly. Usually, comanagement focuses on harvesting activities and the process is confined to that activity alone. However, a number of other functions may be enhanced by the joint action of users and government resource managers at the local level. These include the setting of policy objectives, gathering of data, monitoring and enforcement, estimating resources, making logistical decisions on fishing capacity and limiting fishing effort, conservating marine habitat, regulating market, and research and education (Pinkerton 1989b). No single formula exists to implement a comanagement arrangement to cover these functions. The answer, dependent on country-specific and site-specific conditions, is ultimately a political decision. Nevertheless, stakeholders should be involved in the earliest stages of the policy process, not brought in only in the implementation phase.

These issues are not easily resolved. Each policy bearing on comanagement is embedded in a broader network of laws, policies, and administrative procedures, at both national and local government level, and consequently difficult to change. Government's role in comanagement is to provide enabling legislation to facilitate and support the right to organize and make fisheries management arrangements at the local level, address problems beyond the scope of local arrangements, and provide assistance and services to support the maintenance of local arrangements. Government must ensure that the roles and responsibilities ascribed to user groups are clear, specific, substantive, and permanent. Government administrative and fisheries laws and policies will, in most cases, require restructuring to support decentralization and comanagement. The actual form of comanagement will depend upon the form of government and the political will for decentralization — comanagement cannot succeed in the absence of a favourable policy context.

One must understand that comanagement is an evolving concept with many potential benefits for every group involved. However, to date, successful experiences with comanagement around the world are limited. It should be viewed as an alternative arrangement to be used along with centralized and market governance of fisheries.

8.5 Conditions Affecting the Success of Fisheries Comanagement

Over the last decade, research done at various locations around the world has documented many cases of comanagement and community-based management in fisheries and other natural resource systems (Ostrom 1990, 1992; Pinkerton 1989a; Weinstein 2000). From the results, certain conditions are emerging as central to the chances of developing and sustaining successful comanagement arrangements. These conditions are not absolute or complete: comanagement can occur without meeting all of them. Still, researchers have found that the more of these conditions that are satisfied in a situation or system, the greater is the chance for successful comanagement.

Research is continuing to reveal more about the systems and factors for successful implementation and performance. And even more research is required to establish evaluative criteria for outcomes such as sustainability, equity, and efficiency of fisheries comanagement systems.

We group the conditions contributing to successful comanagement of fisheries in the three categories of contextual variables that Pollnac (1998) identified: 1) supracommunity level, 2) community level, and 3) individual and household level. (See the Appendix for suggested approaches for assessing these variables.)

8.5.1 Supracommunity level

Supracommunity conditions affecting the success of fisheries comanagement are external to the community and include government, legislation, and markets. They can also include demographic factors and technological change. We focus on two conditions:

1. **Legal right to organize:** The fisher group or organization has the legal right to organize and make arrangements related to their needs. The government provides enabling legislation that defines and clarifies local responsibility and authority.
2. **External agents:** External agents (NGOs, academic and research institutions) can expedite the comanagement process. They assist in defining the problem; provide independent advice, ideas and expertise; guide joint problem solving and decision-making; initiate management plans; and advocate appropriate policies.

8.5.2 Community level

Community conditions affecting the success of fisheries comanagement include the local physical and the social environment. The following list describes *preferred conditions*:

3. **Clearly defined boundaries:** The physical boundaries of the area to be managed is distinct so that the fishers group can know them well, and should be based on an ecosystem that fishers can easily observe and understand. The size of the area allows for management with available technology (that is, transportation and communication).
4. **Clearly defined membership:** The individual fishers or households with rights to fish in the bounded fishing area and to participate in area management are clearly defined. The number of fishers or households is not so large that it restricts effective communication and decision-making.
5. **Group cohesion:** The fisher group or organization permanently resides in the area to be managed. The group is highly homogenous in kinship, ethnicity, religion, and fishing gear type. Local ideology, customs, and belief systems create a willingness to deal with collective problems. There is a common understanding of the problem and of alternative strategies and outcomes.

6. **Participation by those affected (inclusivity):** Most individuals affected by the management arrangements are included in the group that makes and can change the arrangements. The same people that collect information on the fisheries make decisions about management arrangements.
7. **Cooperation and leadership at community level:** Fishers are willing and motivated to put time, effort, and money into fisheries management. An individual or core group takes leadership responsibility for the management process.
8. **Leadership:** Local leaders set an example for others to follow, lay out courses of action, and contribute energy and direction to the comanagement process.
9. **Empowerment:** Community members become empowered. Empowerment builds the capacity of individuals and community to increase their social awareness, their autonomy in decision-making, and their self-reliance. Empowerment, which establishes a balance in community power relations, is achieved through education and training. Linked to empowerment are social preparation and value formation toward collective action and the taking on of responsibility for resource management and decision-making.
10. **Property rights over the resource:** Property rights are defined. These, individual or collective, address the legal ownership of a resource and define the mechanisms (economic, administrative, collective) and structures required for an allocation of use rights that will optimize use and ensure conservation of resources and the procedures and means for enforcement.
11. **Appropriate local organizations:** Organizations have clearly defined membership, the legal right to exist, are autonomous from government and political pressures, and represent the majority of resource users in the community.
12. **Adequate financial resources:** Funds are available to support the comanagement process. Sufficient, timely and sustained funding is crucial to the sustainability of comanagement.
13. **Partnerships and partner sense of ownership of the comanagement process:** Active participation of partners in the planning and implementation process is directly related to their sense of ownership and commitment to the comanagement arrangements.
14. **Accountability and transparency:** Business is conducted in a fair and open manner. All partners are accountable for upholding the comanagement agreement.
15. **Strong comanagement institution:** A competent, trusted institution is in place to make decisions and manage conflict. This institution could be created by the comanagement agreement, a committee, or a round table.

8.5.3 Individual Level

Individual decision-making is central to the success of comanagement.

16. **Individual incentive structure:** The success of comanagement hinges on an incentive structure (economic, social, political) that induces individuals to participate. Individuals must expect that the benefits they would derive from participating in and complying with community-based management would exceed the costs of their investments in such activities.
17. **Credible rules and effective enforcement:** An individual must find the rules for management credible and equitable. Vigorous, fair, and sustained enforcement requires the participation of all partners. The motivation to comply with regulations depends on rational decisions in which a person measures the expected benefits of violating the rules against the risk of being apprehended and fined. Furthermore, the state must be willing to use its police powers to support community regulations.

8.6 A Process for Community-Centred Fisheries Comanagement

A process for community-based fisheries comanagement has been developed based on lessons learned from the community-based management programs of NGOs and other institutions around the Asian region. We point out that this is only one possible process of implementing community-based fisheries comanagement — there is no "right" process to develop comanagement. We present the following method as a generic process that can be adapted to meet the conditions and needs of a particular situation. The process described in this chapter is a community-initiated activity that is implemented with the help of an external agent and/or government agency. Another approach, not presented in this chapter, can be described as an externally initiated activity where the external agent or government agency identifies a problem (or problems), then establishes a community-based comanagement project in partnership with the community.

The implementation of community-based fisheries comanagement can be viewed as having three phases: pre-implementation, implementation, and post-implementation (**Figure 8.2**).

8.6.1 Pre-Implementation

The pre-implementation phase of comanagement usually starts when resource users and stakeholders recognize a resource(s) problem that may threaten their livelihood. This is especially true where the resource users are highly dependent on a resource(s), availability of the resource(s) is uncertain or limited, and the users are highly identified with their fishing area. If the resource(s) problem, such as low or no catch, repeatedly recurs; if it exists within a single community of resource users; and if the users are unable or unwilling to move to another fishing area, the resource users are more likely to take action to deal with the problem. Resource users will openly discuss the problem, a process that often leads to consensus building and agreement on a plan of action.

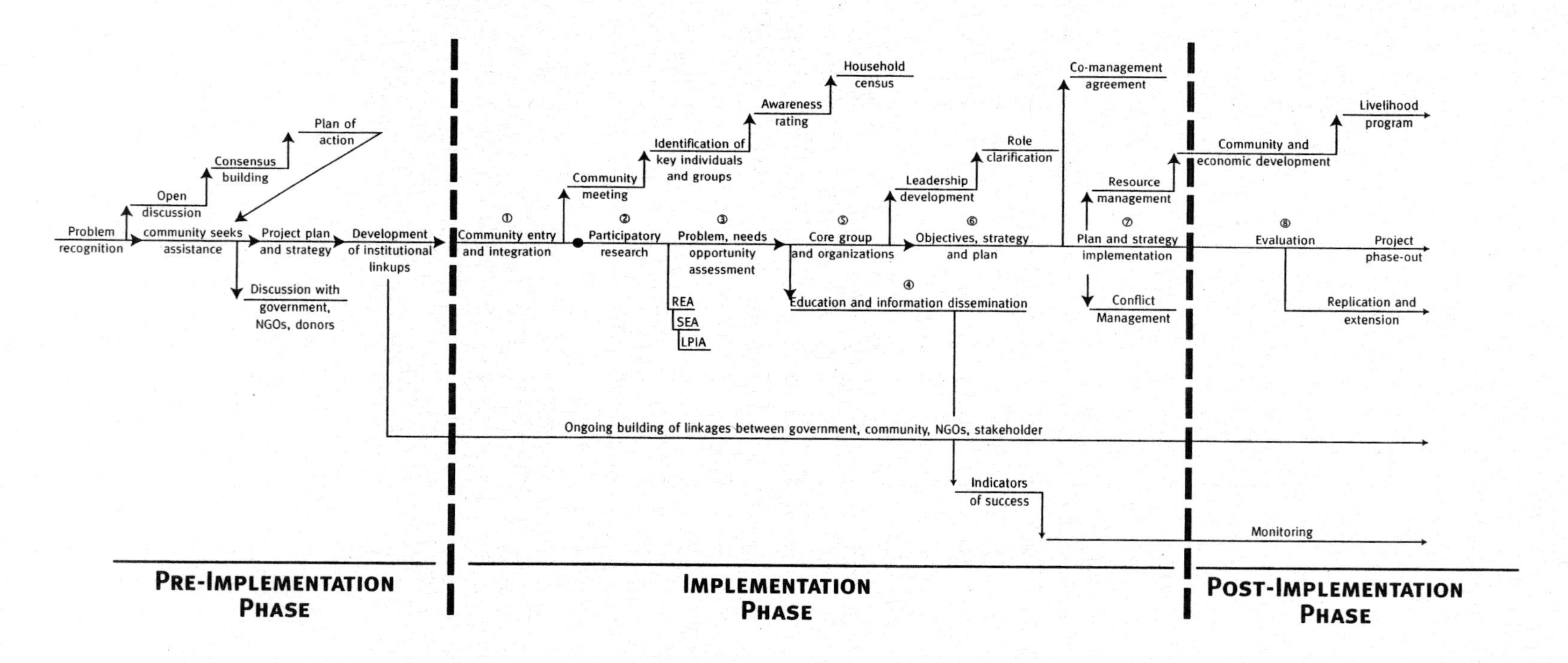

Figure 8.2 A process for community-centred fisheries comanagement.

At this early stage, an enthusiastic individual(s) may step forward as the prime mover(s) of the comanagement process. The resource users may seek assistance from the government or external agents on possible solutions or courses of action to deal with the problem. These outside (of the community) institutions may enter at this point to assist the community, by organizing meetings and providing information, to prepare a preliminary project plan and strategy. A proposal for outside funding of the project may be prepared. Initial approvals for the project may be obtained from different levels of government and local leadership.

At this point, linkages are established and strengthened between resource users and government so that a partnership is developed. A formal or informal agreement for cooperation may be established at this time. The development and strengthening of these linkages and networking, at institutional, group, and personal levels, is a continuous process during the life of the comanagement activities.

In fact, many of the process activities described in this chapter are continuous and overlapping, especially during the implementation phase. The process is dynamic rather than linear, often cyclic as it evolves, and adaptive and pluralistic. The flow diagram in **Figure 8.3** is a simple representation of a complex process. The pre-implementation phase actually flows into the implementation phase.

8.6.2 Implementing Comanagement

The implementation of comanagement has four components: 1) resource management, 2) community and economic development, 3) capability building, and 4) institutional support. Gender, cultural and ethnic issues are emphasized throughout the implementation phase.

The resource management component consists of activities to manage, protect, conserve, rehabilitate, regulate, and enhance marine and coastal resources.

The community and economic development component's purpose is to raise income, improve living standards, and generate employment through alternative and supplemental livelihood development, community social services and infrastructure development, enterprise development, and regional economic development, which includes industrialization.

The capability-building component, aimed at individuals and groups, involves empowerment and participation, education, training, leadership, and organization development.

The institutional support component involves conflict management mechanisms, individual and organizational linkages, interactive learning, legal support, policy development, advocacy and networking, forums for knowledge sharing, power sharing and decision-making, and institution building and strengthening.

The activities in the implementation process of comanagement, described below, are illustrated in **Figure 8.3**.

Activity 1. Community Entry and Integration

Community entry and integration are usually the first steps in implementation. Field workers and community organizers provided by the external agent begin to identify the main stakeholders, those groups and individuals with an interest in comanagement. It is often difficult to determine who is and who is not a legitimate stakeholder and at what level in the comanagement partnership they should be involved. The field workers and community organizers establish initial relationships and credibility with community members, targetting project participants and local leaders at this time. They identify and study the communication and participation structures in the community, including local social structures and power relations, forums for discussion and conflict management, communication barriers by gender and class, and participation in decision-making.

A series of meetings and discussions are held with resource users, stakeholders, and government officials to share the concept and process of community-based comanagement, to begin to develop a consensus on their interests and concerns, and to build awareness about resource protection, management, and rehabilitation. Other activities include identifying key individuals and groups to be involved in comanagement, answering questions about the project, raising awareness about issues, the process and the project, and participating in community activities such as fishing and local events. At this point, it is useful to conduct a feasibility analysis to determine whether a comanagement arrangement would be possible. The legal, political, institutional, economic, and socio-cultural feasibility need to be considered (Borrini-Feyerabend 1996). A household census may be conducted to collect socioeconomic data on the community to identify problems, needs, and opportunities. Community integration of field workers and community organizers can be a long process and requires those workers to have the skills and personality to listen, share, and work with the people of the community on an equal basis.

Activity 2. Participatory Research

Next, participatory research is conducted to collect and analyze baseline data on the community, its people and its natural resources and to generate new knowledge. (See Chapter 5 for details on preliminary appraisal and baseline information.) The baseline data are used in the preparation of development and management plans and strategies, for monitoring and evaluation, and for process documentation. A participatory research process involves the people of the community, working with the researchers, in the collection, analysis, and validation of the output. The participatory research process can also raise awareness and educate community members about their community and natural resources, as well as being useful in the formulation of potential solutions. Participatory research, which is conducted using a mix of scientific and rapid-appraisal methods, includes the collection of traditional and indigenous knowledge.

Participatory research can have three components: 1) resource and ecological assessment (REA), 2) socioeconomic assessment (SEA), and 3) legal, policy, and institutional assessment (LPIA). A REA provides a scientific and technical information base on the coastal and marine resources of the area. It usually includes three interrelated assessments: capture fisheries, coastal habitat (coral reef, mangrove, sea grass), and water quality. A SEA, which provides baseline information and a profile on social, demographic, cultural, and economic characteristics and conditions in the community, includes both stakeholder and conflict analyses. An LPIA profiles the institutional arrangements (formal and informal rights and rules), organizational arrangements, legislation, and policies and programs (internal and external to the community) for coastal resources management. The baseline data also serve as a basis for the future monitoring of the project and for the evaluation of success and impacts.

Activity 3. Community Assessment

The third step is a participatory assessment of opportunities, problems and needs, which is conducted by and with the stakeholders through a series of community meetings, key informant interviews, surveys, and one-on-one discussions. Community members share with each other, as well as with government and external agents, ideas for their community's future and their vision on how to achieve that future. Drawing on the baseline data from the three research assessments, participants assess and discuss the feasibility of developing a comanagement agreement.

Activity 4. Education and Information

Education and training, integral and ongoing activities of the community-based comanagement process, are the main methods of capability building for community members. The external agent usually implements these activities, based on the assessments conducted earlier. The education and training should recognize and build upon the existing experience and knowledge of community members. Education methods, formal and informal, include small-group work, seminars, cross visits, role-playing, radio, video, and fisher-to-fisher sharing of local knowledge. Environmental education is a priority goal of these activites, as is the building of community members' capabilities and confidence so they can make informed choices and decisions about problem articulation, management and development objectives, strategies and plans, and implementation.

Activity 5. Core Groups and Organizations

Because community organizing is the foundation for mobilizing local human resources, community core groups, organizations and leaders are needed to take on the responsibility and authority for management and development activities. These groups, organizations, and leaders, who focus participation, representation, and power sharing in the community, may already exist in the community, may emerge by themselves, or may be newly established. The members of any such group or organization must be willing to take on the responsibility. Existing groups, organizations, and leaders in the community are identified through the stakeholder analysis and LPIA. Various types of organizational structures can become involved, including associations, cooperatives, unions, management councils, and advisory committees. Organizations may be formed at levels ranging from the fisher to the village to the municipal/district to the province/state.

Education and training can empower the core group, organization or leaders, developing their ability to take on management responsibility. Leadership development is an important part of this step, since strong and dedicated leadership is necessary if community-based comanagement is to succeed. Existing community leaders, such as elected officials and senior fishers, play an important role but may be too closely tied to the existing community power structure to be advocates of improved equity. New leaders, often individuals with the motivation but not the means to take on leadership, can invigorate the process and increase its legitimacy. Terms of office for leaders should be short enough to decrease the possibilty of corruption and power grabbing.

Adequate time must be provided for the organizing and leadership development processes. Lack of social preparation is often the cause of project failure. It is during this step that the roles and responsibilities of groups, organizations, leaders, and stakeholders are delineated and clarified. Formal and informal forum(s) for discussion and debate should be established, with stated place, time, and rules for their meetings. The core groups and organizations advocate for support for policies, laws, and local initiatives. Initial consultations and/or planning meetings are held among the partners to develop the comanagement agreement.

Activity 6. Objectives, Strategy and Plan

As a sixth step, the community-level core groups and organizations, working in partnership with other stakeholders and the government, develop a resource management and community development plan whose objectives and strategies include a comanagement agreement. Community members participate in the creation of the plan, validating its drafts along the way. Reaching the comanagement agreement may involve a series of meetings to reach a consensus on its structure and to support negotiation, mediation or arbitration of conflicts. These meetings will involve identifying the key issues, as well as extensive bargaining and compromising in order to reach decisions.

The comanagement agreement must include, specifically stated, a definition of roles, responsibilities, and authority; identification of forums for meetings; conflict management mechanisms; and rule-making procedures. The agreement should be widely circulated to inform and obtain comments from relevant communities and stakeholders. A comanagement body may be established at the end of the process of developing the agreement to represent all the partners. Participants would specify who comanagement body is to represented; what is its mandate; and its level of authority and tasks. Indicators for monitoring and evaluation of the plan are specified. This can be done through a logical framework analysis (LFA) where outputs, activities, verifiable indicators, and means of verification are stated (see Chapter 3).

Since conflicts will inevitably arise, the agreement must contain forms and mechanisms to address and resolve conflict. Conflict management is a process of dialogue and negotiation. A facilitator, mediator, or arbiter may be needed to guide the process toward constructive results. Participants should designate a forum for negotiation and agree on some rules for the process. They may generate and discuss various options for action, formally agreeing on one of those options. The conflict management mechanism should be multi-level to allow for an appeal process.

The financial resources to implement the comanagement plan should be identified early in the process and made available before implementation. If external funding is needed to implement all or part of the plan, this is the time to identify a source and apply for the funds.

It should be noted again that the strengthening of linkages and partnerships and networking between resource users, stakeholders, government, and the external agent is an ongoing and continuous process that extends beyond the implementation phase. The roles and responsibilities of the partners will change and adjust as the community-based comanagement system matures.

Activity 7. Plan Implementation

The activities and interventions of the comanagement plan are implemented through sub-projects. These may be resource management-related, such as marine reserve or sanctuary creation, mangrove reforestation, erosion management, or fishing gear restriction. On the other hand, they may be about community development: such as a water well, a road or livelihood development, such as agriculture, aquaculture, or small business enterprise. The sub-project may be related to institutional support, such as formal recognition of the community organization or passage of a government ordinance legitimizing local institutional arrangements (rights and rules).

Activity 8. Evaluation

Monitoring and evaluation should be central elements of the overall implementation process, although evaluation may be conducted during the post-implementation phase (see Chapter 5 for discussion on monitoring and evaluation). The indicators specified above are used in monitoring and evaluation, both done in a participatory mode. Participatory monitoring allows for interactive learning and a feedback system of success and failure while the project is being implemented. It provides the community and external agents with information, during the life of the project, so they can assess whether activities are progressing as planned, and whether modifications are needed. Participatory evaluation allows those internal and external to the community to evaluate project objectives against results. It allows for planning for the future based on experience. The baseline information collected earlier in the project can be used in the evaluation. The comanagement agreement is also monitored on an ongoing basis, with the partners reviewing the results. Performance indicators may be used to measure progress of the comanagement agreement, process, and implementation (see Chapter 6).

As needed, the responsibilities and rights of partners are clarified, conflicts are managed, and the agreement is enforced — possibly resulting in changes in the agreement or the development of a new agreement (Maine *et al.* 1996).

8.6.3 Post-implementation

At this point, the project, with assistance from an external agent and external funding, is fully taken over by the community and becomes self-sustaining. The post-implementation phase begins. The external agents work through a planned phase-out from the community and the other comanagement partners. The phase-out should be planned and well understood by all to eliminate surprises and minimize problems. Where feasible, people in other communities replicate and extend the results of the project. Fisher-to-fisher training and cross visits can be an effective way to train people in other communities. Project replication and extension can also to enhance the credibility of the community-based comanagement system in the eyes of the community and the comanagement partners, since success often breeds success (White *et al.* 1994b; Johannes 1998b).

8.7 Conclusions

The idea of active participation of local resource users and communities in development and management is not a new one; it has been part of the development process since the 1960s. What is different is governments' increasing commitment to decentralization and comanagement programs. Fisheries comanagement aims at sharing the authority and/or responsibility to manage the fisheries between government, the community of local fishers, and other resource stakeholders. Three important characteristics of comanagement are empowerment, power sharing, and conflict management. Comanagement addresses the crucial management issues of who controls the rights to use the fisheries and who obtains the benefits from those resources.

Comanagement systems that have arisen around the world show promise for addressing many of the requirements for sustainability, equity, and efficiency in small-scale fisheries management. Comanagement is only one strategy, though, and should not be seen as a panacea. Other options include centralized management, territorial use rights, and market regulations, such as rights-based management. The potential advantages and disadvantages of comanagement are well documented. The development of fisheries comanagement systems is neither automatic nor simple, nor is its survival guaranteed. This is a strategy to be determined by political choice; both the community of local fishers and the government have to be restructured, with the ultimate power residing in the government. Fisheries comanagement requires compromise, respect, and trust between all parties.

Chapter 9
New Directions: A Vision for Small-scale Fisheries

9.1 Small-scale fisheries in context

Coastal marine and freshwater resources are under stress. Many of them, along with the ecosystems upon which they depend, are showing signs of collapse as a result of increasing fisheries overexploitation and habitat degradation. At present, roughly 70 percent of fish stocks for which data are available are fully exploited or overfished. The potential impact of the degradation of coastal and marine ecosystems on human health, food security, biodiversity conservation, and local and national economies will be multiplied as coastal populations continue to increase. The pressure on these systems will be multiplied as well.

However, coastal marine and freshwater resources, and the small-scale fisheries that they support, remain critically important. Of the more than 51 million fishers in the world, over 99 percent are small-scale fishers. At present, 95 percent of the world's fishers are from developing countries, producing 58 percent of the 98 million tonnes of annual marine fish catch (FAO 1999). It is estimated that at least 50 million people in developing countries are directly involved in the harvesting and processing of fish and other aquatic products (ICLARM 1999). If we assume an average household size of five persons, then 250 million people in developing countries are directly dependent upon the fishery for food, income and livelihood. Additionally, fish production employs some 150 million people in developing countries in associated sectors such as marketing, boat building, gear making, and bait. (ICLARM 1999).

Approximately one billion people rely on fish as a major source of their food, income and/or livelihood (ICLARM 1999). Assuming that this is accurate, at least 85 percent of these people rely on fish as their major source of protein. The world value of fish as a protein source in 1991 was US$70 billion. This represents approximately one-fifth of the total animal protein consumed on earth (Williams 1996). For 60 percent of populations in developing countries, 40 to 100 percent of their animal protein comes from fish (ICLARM 1999). For all developing countries, fish production in 1995 equaled 60 million tonnes. This figure is equal to all four terrestrial animal protein commodity groups combined (beef and veal, sheep, pig, and poultry) for these countries (IFPRI 1996).

As the information above highlights, the importance of the world's fisheries, and especially the small-scale fisheries, in providing food, income, and livelihood cannot be overemphasized, especially in developing countries. Yet small-scale fisheries have been systematically ignored and marginalized over the years, in both developing and developed countries. In most cases, this was not deliberate but a result of an accumulation of policies and development decisions to "modernize" fisheries. In many countries, the commercial and industrial fisheries have been systematically favoured, often to the detriment of both the small-scale fishers and the fish stocks on which they depend. Major conflicts between the two sectors have been occurring in different parts of the world, from Senegal to Canada, and from Indonesia to Barbados, with resulting threats to food security and local economies, and, in some cases, ecosystem health. In addition, small-scale fishers have been driven out of traditional fishing areas and landing sites as a result of and conflict with tourism, recreational, residential, and industrial development.

Despite all the policy measures in their support, commercial and industrial fisheries have not, in fact, replaced small-scale fisheries. This is true in developed countries such as the United States and Canada; an example is the success and dynamism of the Gulf of Maine lobster fisheries in the United States (Acheson 1988) and the Bay of Fundy inshore fixed-gear fisheries in the Canadian Atlantic (Loucks et al. 1998). It is also true in Asia, Africa, the Caribbean, and Latin America, where small-scale fishers have retained their importance and dynamism. In fact, there are more small-scale fishers today than ever before, producing more protein for human consumption.

The majority of small-scale fisheries have not been well managed. It is now almost universally accepted, for example, that most of the coastal fisheries in Southeast Asia are overfished. With excess capacity, both labour and capital, existing in most coastal small-scale fisheries, most existing fisheries management arrangements have failed to successfully coordinate and restrain fishing capacity and effort in small-scale fisheries and to manage conflict. The management arrangements have not kept pace with the technological ability to exploit the resource or with the incentives driving exploitation and degradation: economic returns, population growth, food, and employment. While governments in many countries are working to attain sustainable development of coastal and marine resources and to improve the social and economic conditions of coastal residents, funds and other resources for these purposes are limited.

As stated in Chapter 1, our emphasis in this book is not the whole of the world's fisheries, with their many and complex problems and opportunities. Rather, our focus is new management directions for small-scale fisheries that operate in nearshore coastal waters using a multitude of fishing gears. These new directions concern the way in which the practical management of these fisheries is approached and implemented. Conventional fisheries science and management has not well served fisheries that are small-scale and based on small stocks (Mahon 1997). Stock assessment-based fishery research and management has been too expensive, too incomplete, too uncertain, and too impractical to address the needs of small-scale fisheries, especially tropical multispecies stocks.

The science of fisheries is not unique in this regard. As Sachs (1999) has observed, scientific research and development, in general, "are overwhelmingly directed at rich-country problems . . . . The international system fails to meet the scientific and technological needs of the world's poorest." Our search for new directions is driven by the need to create an alternative and appropriate science for the poorer and smaller nations of the world, as well as the poorer, resource-dependent regions of the richer states.

However, in searching for new directions for small-scale fisheries management, we are not rejecting conventional science. Rather, we reject the kind of conventional fishery science that has become a relatively unquestioning way to deal with the problems of fishery management. There is a general consensus in many circles that *Reinventing Fisheries Management* (Pitcher et al. 1998) has become a necessity. The mechanical cranking out of stock assessment parameters has lost sight of science as a way of increasing our understanding of the world. As Corkett (1997) characterizes it, application of fishery science has become instrumentalist, involving "the construction of models whose sole role is that of 'positive' prediction, where a scientific theory neither explains nor describes the world." These models come to have a life of their own but, lacking the ability to produce explanatory hypotheses that may be tested against experience, they "are of no more value in the management of the world's fisheries than the magic spells of witch doctors" (Corkett 1997).

But let us not get carried away. If used properly, with due regard to limitations of data and the inherent uncertainties in ecosystem processes, conventional fishery management science possesses many strengths and has a great deal of experience behind it. But one must remember whence it came and for what purpose it was developed; that is, in the service of large, single-stock fisheries using a single type of fishing gear in the North temperate regions of the world. It was developed to deal with the biology, economics, and management of the commercial-scale and large-scale fishers targeting these fisheries. However, it has been shown to be ill-suited to deal with multispecies stocks in nearshore coastal waters that are targeted by multiple types of fishing gears, the characteristics of many small-scale fisheries in developing countries.

The management of such fisheries often requires a different kind of biology to deal with multiple species and ecosystem health. It also requires greater attention to the social context of science (Jasanoff et al. 1997). It requires a very different kind of economics, emphasizing the benefits and costs of not just individual fishing boats and fishing fleets but also fishing households and communities. It requires an understanding of human behaviour and how people use and misuse marine resources. And it also requires a different kind of management regime that goes beyond command-and-control measures.

Conventional fisheries management science does not have the methods in its toolkit to effectively deal with these complexities. If we are going to effectively handle the crisis occurring in many small-scale fisheries in the world, we must go beyond the scope of conventional fishery management. We are not asking that you reject science, for it is the basis of everything we propose. We are, however, asking that you reconsider the conventional, to be creative and innovative, and to consider new ideas and approaches, some of which will require significant changes in the way fisheries are managed and in the way you work. New concepts, tools, methods, and management and conservation strategies are here now. If we are to ensure food security and livelihood of small-scale fishers and conserve the biodiversity of the resources on which they depend, we must consider new directions and take action.

9.2 New directions: a vision for small-scale fisheries

As discussed in Chapter 1, our vision for small-scale fisheries is one in which they are no longer marginalized; where fishers are invited to participate in management decision-making and are empowered to do so; where poverty and food security are not persistent problems; and where the social-ecological system is managed sustainably. It is a vision where small-scale fisheries' contributions to national economies are recognized and given adequate acknowledgement and support. It is a vision that sees the linkages between human and natural systems and recognizes the need for management approaches that address these linkages. It is a vision with a human face and a people focus — fishers and fishing communities. We understand that this is a broad vision, but we feel that it is achievable for many small-scale fisheries. (We also recognize that it is not achievable everywhere.)

Our vision is consistent with the turn-of-the millennium emphasis on horizontal processes such as collaboration, partnership, and community empowerment (Pomeroy 1994b; Haggan 1998). As Max Weber pointed out, bureaucratic authority developed in the early part of the 20th century in the Western world as a condition for organizational effectiveness. This is no longer the case. In the emerging post-traditional social order, to use Anthony Giddens' terminology, conventional resource management and the mid-20th-century bureaucracy starts to disappear. "States can no longer treat their citizens as subjects," as Giddens (1994) puts it. Information produced by specialists can no longer be confined to specific groups but becomes widely available. There is a rise of civil society, including citizen science and citizen action, as in countries such as India (Gadgil et al. 2000); local and traditional knowledge start to play a role in resource management (Johannes 1998a; Berkes 1999). The distinction between science and resource management starts to disappear, as in adaptive management (Walters 1986), and "old bureaucracy starts to disappear, the dinosaur of the post-industrial age" (Giddens 1994).

However, the emerging post-traditional social order is not one in which tradition disappears; far from it. The status and role of tradition changes. But wasn't tradition considered to be in collision with modernity? Wasn't overcoming tradition the main impetus of Enlightment in the first place?

Revival of tradition is in part related to the excesses of "modernization" and the failure of development models that tried to import European and North American approaches and values from the top down. In fisheries, as well as in many other areas of development and management, bottom-up approaches are gradually replacing the top-down style — a change that has been happening since at least the 1980s (for example, Chambers 1983). Thus, local values and priorities are needed to guide development and resource management; they cannot be supplied from the outside as "givens." As well, the revival of tradition is necessary to balance out globalizing influences, providing local pride and a sense of well-being, a social identity and social capital to help people survive in the alienating environment of an increasingly globalized economy.

It is becoming increasingly clear that governments, with their finite resources, cannot solve all the problems that citizens face. The communities will need to take more responsibility for solving local problems. In order to do this, however, they must be given the power and resources to make decisions locally and to take actions to meet local opportunities and problems. They will still need the assistance and support of national government to achieve results.

9.3 New directions: concepts, methods, and tools

A number of promising new and revised management approaches that have emerged in recent years are available for use by managers of small-scale fisheries and for the fishers themselves to use. These include methodological approaches that emphasize fishery and ecosystem management objectives and participatory decision processes rather than focusing, as is usual, on fish stock assessment and population dynamics and paying less attention to the human dimensions of the fishery. Included here are new governance regimes such as community-based management and comanagement that have the potential to address community and economic development as an integral part of fishery resource management. Interdisciplinary and social science methodologies, including versions of logical framework analysis, the use of fishers' local ecological knowledge, and participatory rural appraisal, feature prominently. The management process itself has become more adaptive.

Traditional command-and-control regulatory measures have been supplemented by property rights-based approaches. Ecosystem-based measures, such as marine protected areas, provide alternatives for protecting local fish populations. Integrated approaches seek solutions to the problems in the fishery sector in other economic sectors. Integrated coastal area management may incorporate fisheries issues into the total scheme of coastal economic development using a geographic information system (GIS), thus providing powerful visual information for decision-making and conflict management. Information is becoming more readily available through computer sources. The list of available approaches goes on and on.

It is increasingly important for the fisheries manager to be creative and innovative. There is no blueprint formula for managing a fishery or an ecosystem: each area or community is different. Different approaches will need to be tried and integrated together. There may be failure. There will be learning and adaptation. The community of resource users and the resource manager will need to work together to decide the best combination of approaches to their situation.

Throughout this book, we have given examples of successful cases of the various approaches we discussed. What we have presented is not theory. From the Caribbean to Africa to Asia, people's lives and natural ecosystems are improving. Though this approach is practical and successful, it is not easy. It involves change and risk for all parties, and involves thinking outside the conventional fisheries science and management "box," recognizing that new directions are available.

9.4 NEW DIRECTIONS: HOW YOU GET THERE

In this book, we have proposed that you reconsider how you approach small-scale fisheries management. Adaptation will involve change on the part of all the stakeholders in the process of management: the fishers, their families, resource managers, elected officials, and NGOs. The book presents a great deal of new information: new concepts, methods, and tools for small-scale fisheries management. It can be intimidating and overwhelming — any change is. You don't have to do it all at once. If you do, you will probably fail. No one likes radical change — it involves risk. It is best to start simple and keep it simple. Try one or two of the new directions discussed in this book. Follow this process: experiment; adapt; learn; share; set an objective and select an approach (or approaches). Try it out. Make changes as needed. Learn from others' activities and share knowledge with them. Not all of these approaches will work in all situations. Communities of small-scale fishers will need to work together and with NGOs and fisheries managers. They will need to take incremental steps to achieve something big — that is, improvements in the resource and in their life.

The reality is that we have to do something. Most small-scale fisheries and fishers around the world are in crisis. Since the current management approaches are not effective, trying something new may be better than maintaining the status quo.

What do you have to do? You will have to learn and think about these new concepts and techniques. Yes, it will take some work. It will involve study and discussion with others. As difficult as it may be, you will have to put aside your biases, whatever they may be about: the behaviour of fishers, the behaviour of managers, scientific superiority, the corruption of government. If we are to succeed, we must open our minds and refresh our thinking. This sounds transcendental; new directions often are. But what choice do we have? The future of our marine and coastal resources is at stake. People's lives and futures are at stake. You can make a difference.

Appendix

Questions We Ask in Fishery Management Project Assessment, Monitoring, and Evaluation

Operational definitions of variables

In this Appendix, variables identified in the information needs section of Chapter 5 are described in terms of specific methods that can be used to obtain the information. The methods described are merely examples that should be adapted to local conditions as appropriate. For example, some of the questions, as they are phrased, may be culturally inappropriate in some situations. This can be determined with help from individuals who know the local area. Additionally, all questions should be pretested in the local area before a survey is conducted. It is important to understand, however, that when data is collected as a part of a monitoring and evaluation program, the exact same methods should be used for all sites and at all time periods. This is necessary to ensure comparability when describing differences between sites and time periods.

The section is organized according to the classifications described in Chapter 5: impact, context, and project variables. In some cases, more than one method is described for a variable. Where this is done, the outputs of the different methods are compared in terms of uses, strengths and weaknesses. Examples are provided for the more complicated methods.

1.1 Questions for ultimate impact variables

The "ultimate" impact variables are examined from both a subjective and objective perspective. The subjective perspective involves the community members' perceptions of the "well-being" of their household and the target fish stocks. The objective perspective involves determining the state (measuring, either quantitatively or qualitatively) of the "well-being" of the households in the community and the target fish stocks. For comparative purposes, these variables are evaluated at all stages of the fishery management project.

1.1.1 Participants' perceptions

It can be argued that the subjective perspective is as important as the objective; hence, they both should be evaluated. Participation in, as well as sustainability of, a project is based in large part on participants' reactions to the project. In turn, these reactions are based on user perceptions of the "well-being" of their household, the community and the target fish stocks, which are not always in accord with objective, quantifiable evidence. For example, in cases where the natural environment is degrading, a resource scientist would label a steady state as an improvement, viewing it positively. Fishers, on the other hand, may not perceive the "steady state" as an improvement. Likewise, if restrictions are placed on fishing areas or methods, fishers may view decreased catches as an indication that fisheries management project activities are not improving the natural environment. Hence, if there is an interest in understanding participation in as well as sustainability of fisheries management projects, it is essential to understand perceptions of the "well-being" of the households in the community and the target fish stocks. Perceptions of these indicators may explain some of the variance in long-term, as well as short-term, project success. Impact indicators suggested for this study are:

1. Perceived well-being of the target fish stocks;
2. Perceived well-being of the fisher's household;
3. Perceived household income;
4. Perceived control over the resources;
5. Perceived ability to participate in community affairs;
6. Perceived degree of compliance with fishery management.

Ideally, the method used will take advantage of the human ability to make graded ordinal judgments concerning both subjective and objective phenomena. For example, one has the ability to evaluate real-world objects in terms of some attribute such as size and not merely make the judgment that one is larger than the other but see that one is a little larger, much larger, etc. Human behaviour is based on graded ordinal judgments, not simply a dichotomous judgment of presence or absence. For example, a person is more likely to take action if they perceive that an activity will benefit them "greatly" in contrast to "just a little." This refined level of measurement allows one to make more refined judgments concerning fisheries management project impacts, as well as permitting use of more powerful non-parametric statistical techniques to determine relationships between perceived impacts and potential predictor variables. Several techniques can be used to evaluate individual perceptions of the above indicators.

First, one could be requested to express degree of satisfaction or dissatisfaction along a 7- (or other) point scale. This procedure would involve informing the subjects that they will be requested to report how satisfied or dissatisfied they are with certain aspects of their environment and living conditions. Then, for each topic, the subjects will be asked to respond whether they are satisfied, dissatisfied, or neither. If they respond "satisfied," they will be asked if they are very satisfied, satisfied, or just a little satisfied. The same procedure would be applied to a "dissatisfied" response. Including the "neither" or neutral response, this results in a 7-point scale, with 1 indicating very dissatisfied and 7 very satisfied. Respondents would be requested to make these judgments for today and some time in the past (for example, 10 years before the baseline, for the baseline or the time of initiation of a management project for post-evaluation) to obtain trends. This technique might prove to be unreliable for uncovering minor changes between time periods due to the size of the categories used.

Another technique uses a visual, self-anchoring, ladder-like scale that allows for making finer ordinal judgments, places less demand on informant memory, and can be administered more rapidly. Using this technique, the subject is shown a ladder-like diagram with 15 (or whatever number) steps. The subject is told that the first step represents the worst possible situation. For example, with respect to fish stocks, the first step might indicate that none of the target fish are in the water. The highest step could be described as indicating so many fish that the fisher can easily catch as many fish as necessary in a short period of time. The subject would then be asked where on this ladder (ruler, scale, whatever is appropriate for the subjects involved) the local area is today (this is the self-anchoring aspect of the scale). The subject would then be asked to indicate where it was during some past period of time (for example, 10 years before the baseline, for the baseline or the time of initiation of a management project for post-evaluation). It can be argued that such scales can be treated as "quasi-metric," permitting the use of parametric statistics with fewer reservations than with the previously discussed technique.

The two techniques yield information that is similar but subject to slightly different interpretations. For example, a position on the self-anchoring scale does not necessarily indicate satisfaction or dissatisfaction, and we might be in error if we interpret a scale value above the mid-point as indicating individual satisfaction. Likewise, satisfaction with an attribute (for example, income) does not tell us where in the perceived range of income the individual places him/herself. The self-anchoring scale, however, will be both easier to administer and more sensitive to the changes we want to evaluate.

Perceived changes are only one aspect of the evaluation. It is also important to determine individual explanations for the changes. This can be achieved by asking the subject why a given change has occurred. This open-ended type of question can provide valuable insights related to community perceptions of factors influencing perceived changes.

1.1.2 Objective evaluations

Using "well-being" as an objective assessment of a human community can be a complex, expensive process. Variables often mentioned as indicators include income, health and nutrition status, housing and education. For example, health and nutrition status are notoriously difficult to assess in developing country contexts. Mounting evidence questions the reliability of informant recall as a method to obtain such information (cf. Ricci et al. 1995; Bernard et al. 1984), and employment of biological anthropological techniques, such as skin fold measurements, would be both expensive and time consuming. Income information is also difficult to obtain, especially among fishers whose day-to-day catches vary so extensively that informant recall is highly unreliable (cf. Stevenson, Pollnac and Logan 1982). This results in the use of complicated techniques asking for income on good, average and bad days, then trying to obtain information to calculate the approximate number of each type of day per fishing season, etc., then calculating an estimated income. More accurate information can be obtained from landing statistics, but they are rarely collected and frequently unreliable. Frequently, the estimate is made based on an "average" (as variously understood by individual fishers) fishing trip, which tells us little. Finally, estimates of income are further complicated by the occupational multiplicity that characterizes rural areas in developing countries. Education and housing are a bit easier to assess. Housing is frequently assessed using some type of material style of life scale composed of house construction and furnishings attributes (cf. Pollnac et al. 1989).

Sometimes these highly interrelated variables are combined in some fashion and referred to as "quality of life." A traditional, single-item indicator of quality of life is infant mortality rate. This is a fairly good measure of general nutrition and health care, as well as indicators concerning satisfaction of some basic human needs and related to income and education. Newland (1981:5) writes that "no cold statistic expresses more eloquently the differences between a society of sufficiency and a society of deprivation than the infant mortality rate." Secondary sources might provide this information for the fisheries and coastal management project target area, but it is most likely aggregated for some larger area; hence, inappropriate for estimating project impacts. Regional health services may have the disaggregated data that could be used to calculate an index for the fisheries and coastal management project context but the population might be so small that an excessively long series of data would be required to arrive at a reliable infant mortality rate, suggesting that attempts to use the rate to evaluate changes over a period of several years would be inappropriate.

1.2 Questions for Intermediate Impact Variables

Most of the indicators can be assessed using project reports, information from key informants, and visual inspection. Several indicators will require use of survey methodology, and the description is preceded with "social survey question."

1. Make a list of the project's intermediate objectives.
2. Determine degree of achievement of objectives.
3. Evaluate current status of achievements:

Material Aspects

For each material objective, determine the manner in which achievements have been sustained; for example, are the artificial reefs still in place and being used as described in objectives? Are the mangrove plantings (habitat enhancement) still growing and being cared for, as in objectives?

Institutional and Organizational Aspects

3a. Use rights *developed from project efforts* are evaluated using the categories listed for evaluation of use rights as a contextual variable.

--a1. In terms of the relevant resource, are there any restrictions concerning who has rights to harvest the resource?

--a2. Are the rights restricted to a) an area or region? b) a particular species? c) use of a particular gear? d) certain recreational activities? e) other (specify)?

--a3. If yes, is there written legislation concerning these rights or are the rights based on an informal agreement?

--a4. Who has the right of access and who is excluded?

--a5. Describe the boundaries in terms of distinctness.

--a6. Is it possible to transfer the access rights (for example, by inheritance, by selling them or giving them away)?

--a7. How would one be caught if they break the access rule?

--a8. How would they be punished?

3b. Are adequate provisions made to monitor compliance with rules (for example, availability of guards, patrol boats as needed)?

--b1. Is the state willing to use police powers to support community regulations?

3c. Are the rules being obeyed? (This question should be posed to all key informants, including some users, enforcement personnel, government representatives, project personnel, etc.)

3d. What types of user participation are provided for in terms of rule making, rule modification, monitoring and enforcement?

3e. How are conflicts between resource users resolved?

3f. Are individuals or core groups (for example, user associations) developed for leadership responsibility in the fishery management project?

3g. If yes, for each association or core individuals: Has education and training for empowerment (for example, social preparation and value formation toward collective action, as well as responsibility for resource management and decision-making) been carried out? What are the people's or groups' functions? Have they demonstrated an ability to carry out these functions? What is the leadership (representativeness and quality)? What is the number of members and the current trend in membership (for associations)? From what area are members drawn? What are social categories of members (occupational categories)? Is the association included as a member of a group of associations? If yes, describe the structure of the grouping of associations. **Social survey questions**: Do you feel that the (association or core individuals) are fair in the management decisions they make? Do you understand how they make their decisions?

3h. **Social survey question**: Why is there a rule about ....? (fill in with regulation developed as a part of the fishery management project. Repeat for each fishery management project–developed resource management regulation).

3i. **Social survey question:** What do you think of this regulation? (appended to question 3h)

3j. **Social survey question:** (if there is a user association involved in management) What are the benefits of belonging to the ...... association?

3k. **Social survey question:** Do you feel that you will be better off as a result of the new management rules?

3l. **Social survey question:** Do you feel that everyone will benefit the same from the new management rules, or will some benefit more?

--ll. If some will benefit more, who and why?

1.3 Questions for Supra-Community Level Context Variables

1.3.1 Enabling Legislation

Operationalization of "enabling legislation" must take into account all the important aspects of legislation discussed in Section 3.2.1 of Chapter 5. Published legislation must be evaluated in terms of: 1) allowing formation of resource-user groups, 2) authorization of resource-user groups to define boundaries for exclusive access, 3) provisions for tenure security, 4) provision of general guidelines within which resource-user groups can devise and implement locally appropriate management rules, 5) provision for recognition and formalization of traditional or informal management systems where they exist, 6) provision of supportative administrative structures for comanagement functions, and 7) provision for participation of user groups in developing surveillance and enforcement methods. The simplest operationalization would be seven basic questions, answered either yes or no:

1. Does the legislation allow formation of user groups?
2. Does the legislation authorize user groups to define boundaries for their exclusive access?
3. Does the legislation provide or allow for the development of mechanisms guaranteeing security of tenure?
4. Does the legislation provide general guidelines within which user groups can devise and legally implement locally appropriate management rules?
5. Does the legislation provide for recognition and formalization of traditional or informal management systems, where they exist?
6. Does the legislation provide for supportative administrative structures for comanagement functions such as:
   a. resource monitoring?
   b. surveillance?
   c. enforcement?
   d. conflict resolution?
   e. information?
7. Does the legislation provide for participation of user groups in developing and implementing surveillance and enforcement methods?

Each of the above questions can be treated as an independent variable. They could also be summed, resulting in a composite measure of "enabling legislation." If they are summed, a procedure would have to be developed to account for the five categories of supportative administrative structures; for example, weight them 0.2 each, and so on. Alternately, each of the six items could be analyzed at a more complex level; for example, analyzing exclusive access and boundary types (cf. Pollnac 1984, 1998). This level of detail does not seem to be necessary for the type of evaluation proposed here.

1.3.2 Supracommunity institutions

The first step is to identify all supracommunity level institutions and organizations that are somehow linked with or impact local fisheries management. Each should be classified according to level of operation, objectives, and impact/influence at the local level. Question format is as follows:

1. What are the names of organizations or institutions outside your community that influence fisheries management?
2. For each institution, identify a) level of operation (for example, national, regional), b) type of link with local organizations, if any (for example, horizontal, vertical, nested), c) objectives (from officer of institution or written objectives), and d) local-level impact (from local key informants).
3. Do government officials actively support local organizations and institutional arrangements? (**Question to key informant**: When is the last time a government official visited the community to ask about or assist with [fisheries resource activities]?)

More complex characterizations of these external institutions (for example, history, resources) could be identified (cf. Pido et al. 1996), but this does not seem necessary for the type of evaluation proposed.

1.3.3 Supracommunity markets

Operationalization of the supracommunity market variable would include questions on external market structure, availability, location, and stability (including consideration of recent changes). Format for the questions is as follows:

1. In terms of markets, what is the distribution of the relevant fishery product?
   a. local percentage
   b. regional percentage
   c. national percentage
   d. international percentage
2. Have there been any recent changes in demand, structure, price or stability in the market?
   If yes, what?

1.3.4 Other supracommunity "shocks" to the system

Questions concerning introduction of new, non-project-related technology (used by locals or "outsiders" now exploiting resources in the area) can be determined through interview of key informants. Other "shocks" (war, political instability, earthquake, typhoon, etc.) can be determined from secondary literature and key informants.

1. Have any new technologies been used in the area since (the year the project began)? If yes, what? By whom?
2. Are outsiders extracting local resources? If yes, where from? What are they doing?
3. Other post-project-implementation "shocks":
   a) natural disasters (for example, typhoon, drought, flood, earthquake).
   b) political or economic restructuring (for example, new tourism or other industry that affects the coast, political instability, war).

1.4 Questions for community-level context variables

1.4.1 Perceived crisis in resource depletion

This assesses local leaders' perceptions of resource depletion (for example, mayor, president of fishers' association, and other opinion leaders such as high-liner fishers). The information will be gathered from selected key informants, in response to the following question:

At the present time the (relevant fishery resource) is:
a) in very good shape.
b) in good shape.
c) in neither good nor bad shape.
d) in bad shape.
e) in very bad shape.

This variable will also be treated as an individual-level variable, with the question posed to a sample of resource users. In terms of the community-level variable, modal response and distribution of key informants' responses will be used as the indicators. Distribution of responses is also used as an indicator since the degree of consensus among opinion leaders may influence other users' attitudes and ultimately the success of the fisheries management project.

1.4.2 Target species composition and distribution

We are interested only in limited aspects of species identification and distribution. The distribution of a target species obviously influences fishing localities, and if the species are concentrated in a relatively small area (for example, around limited reef areas, in several deep holes, on a small expanse of mudflat), increases in numbers of harvesters or effort will rapidly result in conflict-generating pressure on the resource.

Species composition and distribution are also mediating variables with respect to the influences of levels of commercialization and technological changes. For example, if a highly localized species becomes commercially valuable, or if technological changes result in better and more efficient access to a species, fisheries and coastal management project efforts could be thwarted (cf. Pollnac 1984).

Several techniques can be used for determining important species harvested. The simplest is the use of secondary information where available. Many countries collect some form of fisheries statistics, and these could be reviewed as potential sources of information. Since fishery and other coastal resource statistics are notoriously difficult to collect, usually focus only on commercially significant species, and are frequently available only as grouped data for a larger region, they can be misleading, so should be used only as material supplementary to information from key informants unless recent landing statistics, collected in the project community, are available.

Researchers can obtain information from key informants by requesting different categories of users of the relevant resource to list, in order of importance, the most important species they capture for income and home consumption (on separate lists). Information must be gathered from the different categories of users of the relevant resource because users of different gear types frequently target different species. This type of information should be collected from at least five representatives of each major gear type (that is, gear types used by at least one-fourth of the resource exploiters). Commercial significance can be cross-checked through interviews with key informants from the marketing sector. Informants should be asked where they harvest each important type. Ideally, and if appropriate, resource mapping techniques can be employed. Questions can take the following form:

1. Taking the entire year into consideration, what are the most important species harvested in terms of income? (list in order of importance)
2. Taking the entire year into consideration, what are the most important species harvested for home consumption? (list in order of importance)
3. For each type listed, where are they are harvested?
4. For each listed, what changes in abundance have you noticed?

Researchers should take some important considerations into account when collecting this type of information. First, the importance of different species changes according to season. It is important to obtain the ranking for the entire year, not just the season that the interview is taking place. Second, it is important to determine the taxonomic level of the terms obtained in the ranking exercise. For example, important fish species are frequently more highly differentiated in local languages than less important species. Nevertheless, a certain term frequently groups the different types into a larger, more inclusive category. Most often, this more inclusive term will be given to outsiders in response to a question about important species when, in fact, only a few of that type are important. All the types *might* be important, but the researcher must determine this. Hence, it is important to verify if each type listed is a general category or a specific type. An interviewer who does not know the local terms can use a picture book, with the help of local fishers, to make scientific identifications.

1.4.3 Fishing methods and associated target species

Obtaining accurate data on fishing activities is not an easy undertaking. The activities are usually seasonal and take place out of sight of land and at all hours of the night and day. Because this is one of the most important types of data collected in the assessment, a multimethod approach is advisable. Due to the importance of this variable, we provide examples along with the description of the methods to be used.

Interviews with officials

In initial interviews with community officials, it might be productive to ask about the presence of different types of coastal activities. A checklist that includes known and anticipated coastal activities for the target region could be prepared and used as a guide for these interviews. If an official appears to be well informed, try to obtain percent distribution (or numbers) of different activities, gear types used, principal species targeted, participants (according to sex and age), seasonality, and distribution and marketing.

Observation

Fishery information obtained during initial interviews with community officials should be verified by other sources of information. One source of verification is enumeration and observations made during a beach walk. General informational beach walks can be conducted at any time. Frequently, the most appropriate time is following the initial interview with community officials. Ask one of the officials if he has time to accompany you to the beach. This act would show that the beach walk has been sanctioned by a higher authority. On the other hand, if the fishers fear or distrust the local authorities, avoid identification with local authority if you can.

Observations of coastal activities can begin at this time, but the enumeration beach walk should be done when most boats are at the shore, during non-fishing times. Boats and gears should be enumerated using local nomenclature.

Enumeration of boats is frequently facilitated by the fact that boats are usually kept on the beach side of dwellings or moored in the water. Nevertheless, the fieldworker must investigate the land side of dwellings as well. In one area of the Philippines where one of the authors worked, fishers from the hills behind the village kept their small outrigger boats suspended on forked sticks in a boat "parking lot" on the land side of dwellings along the coast.

In some cases it will be impossible to find a time when all boats are ashore. Try to identify the time when most fishers will not be fishing, then use common sense to estimate numbers of boats at sea. Ask what types of boats are out at the time of the enumeration, and determine where they fish. If they fish the inshore area, within sight of land, an estimate can be made by counting the boats at sea. If fishers from another community fish in the same area, ask local fishers for an estimate of numbers. Sometimes the number of boats at sea can be estimated according to shoreside evidence such as mooring buoys, logs used as rollers to bring the boat ashore and on which the boat is stored while ashore, and tracks made in the sand when the boat is pushed to the water.

Some gears are relatively difficult to observe. Large gears (for example, nets) are often stored in the boat, but this varies from region to region. In areas where fishers are concerned with theft, or if the net can deteriorate when exposed to sunlight for long periods, the nets may be stored indoors or under tarps. Smaller gears (for example, small nets, hook and line, spear guns) are almost always stored in a small shed, or in or adjacent to the fisher's dwelling. Obviously, simple counting of gear types is not usually possible. Although a simple counting of gears on the beach would result in unreliable information, observation of stored gears, gears in boats, and gears deployed can be used as starting points for questions about the numbers of various types of gears.

Another significant observational method to identify techniques and species is to find out the landing places and times and be there to observe landings. Though this technique may suffer from the seasonality problem, it benefits from direct observation of behaviour.

Key informants

Preliminary preparation of a taxonomy of species, boats and gears will greatly facilitate the acquisition of accurate information from key informants. The importance of knowing the target region's fishery vocabulary cannot be over emphasized. Raymond Firth, an economic anthropologist with extensive experience in fishing communities writes:

> Furnished with the right word, one can get a direct answer to a question or understand a situation at once; without it, however correct one's speech may be grammatically, one may often puzzle one's informant or be reduced to giving and receiving laborious explanations which often irritate the person one is talking to (Firth 1966:358).

The most effective way to prepare a preliminary taxonomy is to spend several days walking the waterfront, observing fish landings and asking for names of every boat, gear, and fish type observed. Pay attention to minor variations between boat and gear types. The differences may signify not only a different type of vessel or gear (which will probably have a different name) but also different fishing methods and target species.

Walking around and asking questions is a good technique for identifying knowledgeable individuals who are willing to provide useful information. For example, if you observe fishers beating the water with sticks, ask a nearby fisher what they are doing. He will probably respond that they are scaring fish into their net. Then ask, what kind of a net? What kind of fish? Is there a name for that kind of fishing? What is the approximate size of an average and a good catch? What is the work-group size associated with the technique and how many groups are there? Are there special times of the year when the technique is used? Are there other ways to use that kind of net? If yes, what are the names of the other techniques used to deploy the net? What are the target species of each named technique? What is the work group size associated with each technique and how many groups are associated with each technique? What are the seasons for use of the technique?

Most producers like to talk about their activities to someone who is sincerely interested as well as knowledgeable. Be prepared to answer questions concerning boats and methods you have seen elsewhere and to make comments about similar techniques where appropriate. Such comments stimulate informants to provide even more detailed information in the hopes that they can learn something from the interviewer.

A useful technique for stimulating interest is the use of a good fish identification guidebook with coloured pictures. Fishermen love to look at fish and talk about them. The process usually draws a small crowd of participants — fishers, fish sellers, and children — who will provide local names as well as names for techniques used to capture the fish. These taxonomies will not be simple. This is especially true for multispecies tropical fisheries where local taxonomies will name several hundred species (see Pollnac 1998).

Box A1.1 Example of the description of a fishery type in North Sulawesi.

The *giop* provides a great deal of employment and income in Tumbak, North Sulawesi, Indonesia. There are specific seasons associated with this gear. January through April (approximate), when the north wind blows, is the *roa* (halfbeak, *Hemirhamphus sp.*) season. May through August (approximate), when the south wind blows, is the mackerel (*deho*) and skipjack (*cakalang*) season. September through December is low season, but one can still catch something then.

During the low period, the average catch is approximately two boxes a trip but sometimes goes up to 10. A box holds approximately 70 kg of fish. A good catch is 10 to 15 boxes of skipjack or mackerel. For *roa*, high catches are up to 25 boxes (20 000 pieces), but a normal catch is 1/10th that amount, or 2 000 pieces. When the season is right, good catches can be had every three to four days. *Giop* fishers report that in the past there were more fish to catch but because the price was so low (because of poor marketing) they did not catch as many. In the past, they report, they could fish in front of Tumbak; now they have to go out farther. They also noted that in the past they could fish with oars but now they need a motor to get to the good fishing area. They claim that there are fewer fish today and they do not know why. They have to look more and spend more time, but with motors they can do it.

Source: adapted from Pollnac et al. 1997

Survey

More accurate and reliable information concerning gears and their target species, along with approximations of catches, can be obtained by means of a household survey. Fishers can be asked to list the gears they use, the predominant species they capture, and average amounts of harvest. Though this is time consuming it increases the reliability of information acquired. **Boxes A1.1** and **A1.2** illustrate presentation of distribution of fishing methods and a description of species and approximate catch associated with the *giop* (traditional seine net) in two villages in North Sulawesi, Indonesia.

BOX A1.2 EXAMPLE OF INFORMATION ON DISTRIBUTION OF GEAR TYPES IN A FISHING VILLAGE IN NORTH SULAWESI.

A survey interview form was administered to a random sample of 41 households in Bentenan, and 40 in Tumbak, North Sulawesi, Indonesia. If fishing was a reported household activity, the respondent was requested to indicate the gears used and the three most important species harvested with each gear. Results of the survey are in the table below.

Percent distribution of gear types used among sample households involved in the capture fishery

Gear Type	Tumbak	Bentenan
Hand line	54	78
Gill net	20	11
Seine net (*giop*)	43	04
Purse seine (*pajeko*)	—	19
Seine net (*tagaho*)	26	—
Shark net	03	04
Dip net (*sibu-sibu*)	—	04
Speargun (*jubi*)	09	—
Harpoon (*tombak*)	03	04
Light boat	—	11
Compressor	03	—

Note: Columns may sum to more than 100 percent due to the fact that one household may use more than one gear.

Source: adapted from Pollnac et al. 1997

1.4.4 Environmental features influencing boundary definition

An area with numerous reference points, such as a rocky, much-indented coastline with rocks jutting out of the water, small offshore islands and reefs, and numerous shore-side features would facilitate demarcation of use-right areas. In contrast, a long, broad, sandy beach with a relatively even, sand or mud bottom, gradual drop-off and no offshore islands would create difficulties in boundary demarcation. Information to be obtained is as follows:

1. Verbally describe the reference points of the project area.
2. Make an ordinal evaluation of the difficulty involved in defining boundaries for the relevant parts of the project area.

1.4.5 Level of community development

A scale of level of community development based on availability of basic services provides a general context for the fisheries and coastal management project. Suggested items for the scale are as follows:

1. hospital
2. medical clinic
3. resident doctor
4. resident dentist
5. secondary school
6. primary school
7. public water supply
8. water piped to homes
9. sewer pipes or canal
10. sewage treatment facility
11. septic or settling tanks
12. electricity service
13. telephone service
14. food market
15. drugstore
16. hotel or inn
17. restaurant
18. gas station
19. public transportation
20. hardtop road access
21. banking services

Information for the scale can be obtained through a combination of observation and key informants. Each item will be checked as present or absent, then a summary scale is constructed by summing the items. After sufficient numbers of evaluations are conducted, the items can be subjected to some sort of scale analysis (for example, factor analysis) for more sophisticated investigations of the relationship between community development and success of the fisheries management project.

1.4.6 Degree of socioeconomic and cultural homogeneity

There are several types of community homogeneity/heterogeneity. Socioeconomic homogeneity can be evaluated through an examination of income distribution, but reliable surveys of income are difficult and time-consuming; hence, it would be unrealistic to conduct such a survey for the purposes of this evaluation. Socioeconomic heterogeneity can also be evaluated through an examination of distribution of occupations, a type of data discussed above. If the entire community is composed of fishers, it would be occupationally homogeneous. It would be less homogeneous if there were variation between households with respect to source of income; hence, number of distinct occupations could be considered as a measure of occupational heterogeneity. This figure would have to be qualified with a statement concerning the relative concentration of households in any single occupation. For example, a statement that seven distinct occupations are present but 70 percent of the families depend on farming would indicate a lower level of heterogeneity than seven occupations with 10 percent of the population in each. Cultural heterogeneity can be measured based on the number of religious and ethnic groups in the community but a similar qualification concerning relative percentages in each group would also have to be presented.

1. Identify the total number of occupations present in the community and determine the percentage of households engaged in each occupation (data collected as part of occupation structure above).
2. Identify the total number of ethnic groups in the community and the percentage of population represented by each ethnic group.
3. Identify the total number of religious groups in the community and the percentage of population represented by each religious group.

1.4.7 Tradition of cooperation and collective action

An indicator of a tradition of cooperation and collective action would be the number and duration of local groups (associations, cooperatives and so on) formed for cooperative or collective action. Construct a list of groups active at some time during the past 10 years. The groups indicating cooperative or collective action should be identified through an examination of reported group activities. Determine the number of years that each of these groups was active. A summary figure, "group-years," calculated by multiplying the number of groups times the number of years active, could be used as a relative measure of tradition of cooperation and collective action.

1. Identify groups indicating cooperative or collective action that were active sometime during the past 10 years.
2. Describe actions that each group undertook.
3. Determine the total number of years each of these groups was active.

1.4.8 Population and population changes

Population figures for the fisheries and coastal management project area can be obtained from census material. These figures are usually available at 10-year intervals. The current population can be obtained from the most recent census, and the rate (average annual change) can be calculated from the 10-year change. If local censuses are conducted at more frequent intervals, a variable can be added to qualify the average rate. This variable would indicate if the rate was more or less rapid during the preceeding few years.

1. Obtain most recent census figures and figures from 10 years in the past for fisheries and coastal management project area.

1.4.9 Degree of integration into national economy and political system

This evaluation is based on summed ordinal evaluations of links (market, transportation, communication, political) to the larger society.

1. Market:
--no links = 0
--low level of links (some specialty products, such as dried sharkfin are collected by a few buyers who occasionally visit the community) =1
--medium level of links (limited amounts of fish are processed [iced, smoked, and so on] and shipped daily in small quantities to nearby urban areas, either by public transportation or small, privately owned trucks) = 2
--high level of links (most of the catch is processed in processing facilities and trucked to urban areas and/or air freighted to more distant areas) = 3
2. Transportation:
--no links = 0
--low level of links (unimproved roads, seasonally impassable, with no more than a few small public transportation vehicles, such as public taxis or pickup trucks with seats in the back passing through daily) = 1
--medium level of links (improved roads, several daily links via vans or small buses to transportation centres with links to the rest of the country) = 2
--high level of links (good roads, frequent bus departures for other areas, frequent local transportation) = 3
3. Communication:
--none = 0
--low (no telephones, only telegraph or radio links) = 1
--medium (few telephones, usually only in mayor's office or army post) = 2
--high (many telephones, both private and public) = 3
4. Political:
--none (no governor- or congress-level politicians visit the area) = 0
--low (politicians visit rarely, less than once per year) = 1
--medium (politicians visit at least once a year) = 2
--high (politicians visit more than once a year) = 3

The ordinal values for each component of this variable can be considered separately, then summed into a total figure to represent degree of integration into the national economy and political system.

1.4.10 Occupation structure: degree of dependence on and level of commercialization of fishery resource

This is a description of the income-generating activities that depend on fishery resources (for example, fishing, aquaculture). Briefly describe other income-generating activities in terms of category (for example, farming, mining, forestry, industry, service) and total employment. Descriptions of fishery-dependent activities will be detailed, including the number of firms, their size and the number of individuals employed in each activity. The following types of information can be obtained from secondary information (for example, local census) or key informants:

1. For fishery-dependent activities: List all activities by type: fishing, aquaculture, fish processing, fish marketing (wholesale and retail). For each type, identify a) number of firms, b) size of firms, c) number of individuals they employ.
2. For activities not dependent on fisheries: List all activities by category: farming, forestry, industry, service. For each category, identify the number of participants.
3. Have several local key informants (for example, mayor's office, chamber of commerce members) rank order all the activities to show their relative importance to the local economy.
4. Derived the community's degree of dependence on fishery resources from the number of participants in coastal activities and from the relative ranking described in question 3.

1.4.11 Local political organization

This involves a basic description of formal, governmental political organization at the local (municipal or district level) in terms of boundaries, positions, duties, and performance. Techniques for conflict resolution will also be described in terms of roles and positions of involved individuals.

1. Describe local formal political organization.
2. Describe local techniques (formal and informal) for conflict resolution.
3. Map the boundaries of local political units.
4. Using culturally appropriate techniques, determine the quality of local leadership in terms of concern with fishery resources, willingness to cooperate with fishery management process, and perceptions of local people. This very sensitive issue must be carefully approached so as to not turn local leadership against project objectives.

1.4.12 Coastal resource use rights, formal and informal

Determining use rights can be relatively straightforward unless boundaries are illegally maintained (cf. Pollnac 1984). In the straightforward cases, key informants can provide information about: 1) rights to what (for example, habitat, species, gear, recreation), 2) level of formality (for example, based on formal legislation or local agreement), 3) types (for example, whether the access is open, communal, or private), 4) the boundary maintenance system (for example, are boundaries clear and strictly maintained or are they diffuse, with minor transgressions permissible?) (see Acheson 1988), 5) whether and how use rights can be transferred, 6) existence of conflicts in use rights, and 7) types of surveillance and enforcement, if any. This may involve describing the system introduced by a fisheries and coastal management project. If so, determine if any systems existed prior to the fisheries management project and, if so, describe them.

1. Are there or have there ever been any restrictions concerning who has rights to harvest the relevant resource?
2. Are the rights restricted to a) an area or region? b) a particular species? c) use of a particular gear? d) certain recreational activities? e) other? (specify).
3. How long has this system been in effect? If no longer in effect, when was it in effect and for how long?
4. If yes, is there written legislation concerning these rights or are the rights based on an informal agreement?
5. Is there or was there a group or leader to manage and enforce these rights?
6. Who has the right of access; who is excluded?
7. Describe the boundaries in terms of distinctness.
8. Is it possible to transfer the access rights (for example, by inheritance, by selling them or giving them away)?
9. How would one be caught if they break the access rule?
10. How would they be punished?
11. What is the level of compliance (are violations frequent)?

1.4.13 Fishery Resource Management (Other than Use Rights), Formal and Informal

This variable involves a description of local-level fishery resource management efforts, both formal and informal. Information can be obtained from key informants and documents, where available. This may involve describing a system that a fisheries management project introduced. If so, determine if any systems existed prior to that fisheries management project and, if so, describe them. The questions to ask are:

1. Are there, or have there been in the past, any rules concerning harvesting or resource conservation?
2. Are the rules restricted to a) an area or region? b) a particular species? c) use of a particular gear? d) certain recreational activities? e) other?(specify).
3. If yes, how long (was or has been) the system in effect?
4. Is there written legislation or are the rules based on an informal agreement?
5. Is there or was there a group or leader to manage and enforce these rights?
6. How would one be caught if they break a rule?
7. How would they be punished?
8. What is the level of compliance (are violations frequent)?

1.4.14 Local Resource Knowledge

The description of traditional ecological knowledge about fishery resources should include a folk taxonomy of fishery resources, a description of beliefs about important items in taxonomies, and a description of variation in ecological knowledge. Users' ecological knowledge can be determined using ethnographic interview techniques (see Spradley 1969). The first step in acquiring this type of information involves constructing folk taxonomies of fishery resources.

Folk taxonomies for aquatic organisms such as fish and marine invertebrates are best generated using a small group of experienced fishers. Since there is frequently a division of labour according to age, gender (for example, in some societies, females conduct gleaning of invertebrates inshore), or some other criteria, this information must be obtained from representatives of the appropriate subgroups of the community. These subgroups can be identified by means of information gathered as part of the fishery and other aquatic fauna use indicators specified above. The first step is to ask the people to name all the types of fish they know that live on or around the site. One easy way is to ask informants to name organisms they have observed at landing sites and markets. A picture book (colour pictures are best) can also stimulate collection of fish names.

After creating this list, the interviewer can take each name on the list (for example, catfish) and ask if there are any other types of "catfish." List construction will probably take several days, taking up about three hours of the fishers' leisure time on each day. Ideally, the list would be cross-checked with another group, using the same techniques but prompting with items from the first group if they are not named by the second group. Similar methods can be used for other coastal and marine flora and fauna.

Scientific identification of taxonomic items can prove difficult. These lists are frequently surprisingly long. Pollnac (1980), using this technique in an examination of a coastal, small-scale fishery in Costa Rica, elicited 122 named categories of marine fish captured by local fishers. For a coral reef in the Philippines, McManus et al. (1992) list over 500 species of fish associated with a specific reef and Pollnac and Gorospe (1998) list over 250 for a reef in another location. These findings suggest that reef fishers might have more complex taxonomies than the Costa Rican fishers in Pollnac's research. If someone with knowledge of reef fauna and flora taxonomy is present, they can attach the scientific nomenclature to the local name; if not, the researcher should take photographs (or collect samples) for later identification of species he or she is unable to identify. Fish identification books, with colour photographs, can also be used as a supplementary method to link local and scientific names. Photographs also make an excellent stimulus for eliciting names. Where fish change colour and characteristics with age and sex changes, the photographs should include representations of all stages. Some fish also change colour when frightened and/or killed, so these factors have to be taken into account.

In brief, the steps necessary for conducting interviews necessary for generating folk taxonomies include:

1. Identify user groups.
2. Using stimuli such as picture books or organisms in the wild, at landings, and in the market, elicit names.

3. For each name, ask if there are any other types of the type named.
4. Cross-validate information with additional informants.
5. Using fish (shellfish, and so on) identification books, identify fish according to scientific name.
6. Photograph fish types that cannot be identified in the field for identification by experts in the university or fishery department.

For each (or each important) resource, investigators should elicit resource harvester knowledge concerning the resource. For example, for a given type of fish, the investigator should question the harvester (or a group of harvesters, as discussed above for eliciting taxonomies) concerning numbers, locations, mobility patterns, feeding patterns, and reproduction. For each of these information categories, fishers should be queried concerning long-term changes. Reasons for changes should also be determined. Given the species diversity associated with coral reefs, this appears to be a formidable task, but such knowledge will probably be available only for important species: those that the harvesters have been watching, hunting and eviscerating — the ones upon which most of their income depends.

Ethnographic interviewing techniques should be used to obtain this information. A good example of this type of information can be found in Johannes (1981), Lieber (1994) and Pollnac (1980, 1998). Questions that can be used to elicit this type of information for coastal fauna for each organism include:

1. Where is it usually caught?
2. Is it also caught in other areas?
3. Does the area change with time (hour, day, moon, month)?
4. In comparison to other organisms, what is the quantity available?
5. What other organisms are likely to be caught with it?
6. What does it eat?
7. Where and how does it breed?

It is important to note that there will probably be intracultural variation with respect to all aspects of traditional knowledge discussed above (Felt 1994; Berlin 1992; Pollnac 1974). Some of the variation will be related to division of labour in the community, as discussed above, but some will be related to degree of expertise, area of residence, fishing experience, and other factors. The conceptualization of "folk knowledge" as "shared knowledge" implies that care must be taken to not attribute idiosyncratic information as "folk knowledge." This is difficult when using a rapid-appraisal approach, especially given the anti-survey bias held by some ill-informed advocates of rapid rural appraisal. A survey of 10 to 15 fishers concerning key aspects of "folk knowledge" can rapidly identify areas of variability that could be addressed in planning future research for management purposes.

An example of cognitive mapping and variation in local knowledge is presented in Chapter 4, **Box 4.2**. Cognitive mapping is an aspect of local knowledge that is useful in determining knowledge about distribution of fish, breeding areas and so on.

Box A1.3 Example of a partial description of variation in fishers' knowledge of fish types in Atulayan Bay, Philippines.

The fishers in Atulayan Bay, Lagonoy Gulf, the Philippines, have over 260 local names in their taxonomy of fish (Pollnac and Gorospe 1998). Since cultural knowledge is unevenly distributed in any population, one would expect intracultural variability in knowledge associated with a taxonomy as complex as the one used by the fishers of Atulayan Bay. Though adequate investigation of this variability cannot be carried out within the time constraints of rapid appraisal, an example illustrates the difficulties involved. Take the folk generic taxon *linhawan*. In an early stage of our research, an informant was queried concerning *maming*, a Labridae (wrasse). He called it a *linhawan*. He also classified other Labridae (for example, *talad, maming, hipos*) as *linhawan* but included *angol*, the hump-head parrotfish (*Bolbometopon muricatus*, a Scaridae), as a *linhawan*. A review of data collected several days previously, however, indicated that other informants identifying a picture of the hump-head parrotfish as *angol* sometimes use the Tagalog term *mulmol* for *linhawan*. In Tagalog, *mulmol* is identified as Scaridae. These informants noted that *linhawan*, other than *angol*, are classified by colour at the specific rank and gave the examples *linhawang asul* (blue), *puti'* (white), *dilaw* (yellow), and *itim* (black), all of which are Scaridae. Later informants added the folk-specific taxons *buskayan* and *tamumol* to the types of *linhawan* and denied that any of the Labridae are *linhawan*.

1.5 Questions for individual and household-level context variables

Operationalization of the individual-level contextual variables will be based on survey methodology. It will be necessary to obtain the information from a sample of community members in order to make generalizations about the community as a whole and understand the distribution of differential participation in fisheries management projects.

1.5.1 Education

Education is a relatively straightforward variable referring to years of formal education, including technical education. Technical education includes short courses given by fisheries training centres, NGOs, and so on. Information on literacy can be checked using some type of visual technique. For example, the respondent could be asked his or her opinion about a message written on a card, or could be asked to rank a list of items on a card. Because this is a sensitive topic, this method should be used with caution. It would be best if placed at the end of an interview so that embarrassment about literacy will not affect responses to other questions. You could find out about:

1. years of formal education

2. type and duration of other training (for example, short courses, technical school)
3. literacy, which can be checked by means of a visual technique.

1.5.2 Experience

Two types of experience are of interest here: work experience and "group" experience. Work experience focuses on activities dependent on fishery resources, such as fishing, aquaculture, and extractive activities in the mangrove area. It is assumed that all individuals in the sample conduct some type of coastal resource–related activity. Determine the type of activity the person is engaged in at the present time, years that they have been involved in this activity, their previous activities, and years involved in each. "Group" experience refers to their membership in associations, such as fishers' cooperatives. The respondent's age and years of residence in the fisheries management project area are also "experience" variables. Ask:

1. What is your current occupation?
2. How many years have you done this type of work?
3. Have you done any other types of work?
4. For each previous occupation, how many years were you in each? If you have done other work in the past, why did you change to your present occupation?
5. Do you presently belong to some sort of group or association?
   If yes, what? What is the purpose of the group? How long have you belonged?
   If no, did you ever belong to some sort of group or association?
   If yes, what? What was the purpose? Why did you leave the association?
6. What is your age?
7. How long have you lived in this area?

1.5.3 Size and scope of operation

This variable refers to vessel sizes and numbers and ownership status (if a fisher) and pond sizes (if aquaculture), plus estimated production. In all cases, number of coworkers in the operation (boat, pond, harvesting from the mangrove) must be determined. Find out the:

1. number of units (boats, ponds, cages, rafts, and so on)
2. size of each unit
3. number of workers (for example, crew size)
4. production (calculate from easily recalled units such as average daily catch x trips per week x weeks fished per year = estimated annual production).

1.5.4 Technology

This variable identifies aspects of the technology that the respondent uses in his or her fishery resource-based occupation. For a fisher, this would refer to whether the fishing vessel is motorized, types of gear used, and so on. For an aquaculturalist, it would refer to aspects of pond stocking, aeration, fertilization, and so on. For a mangrove harvester, it would refer to whether they use chainsaws, and so on. In general, aim for a brief, basic description of the occupational tools and their deployment.

1. For a fisher: boat type and size; motor (inboard, outboard, HP); gear type (be specific — describe type and deployment, not just "net," etc.); average trip length and average number of trips per month. (If trip length is one day or less, calculate number per month based on number per week. Determine if there are any months when fewer trips are taken.)
2. For an aquaculturalist: use of pumps or gravity for water transfer; aeration; fertilizer; pesticides; material used for enclosures or grow-out locations (for example, cages, ponds).
3. For others: descriptions analogous to the above two.

1.5.5 Cultural values

One way to evaluate cultural values is to infer them from responses to attitude questions. The following attitude questions refer to values significant to fisheries management projects.

1. Do you think that the people in this community can work together to solve community problems (digging a well, clearing a road, and so on)?
2. Do you think fishermen could work together to solve a problem in the fishery such as illegal fishing (blast fishing, use of poisons, and so on)?
3. Should the government, the fishers, or both work together to solve a problem in the fishery?

1.5.6 Job satisfaction

There are complicated and there are simple ways to measure job satisfaction (cf. Pollnac and Poggie 1988). Although the more complicated provide detailed information about the non-pecuniary aspects of an occupation, the best indicator of job satisfaction might be the response to this question: "If you had your life to live over, would you still become a . . . ? (compare with Robinson et al. 1969). Take care to insure that this question is culturally appropriate in your situation. If it is not, eliminated it and use only the other questions suggested below. Meaningful, qualitative information can be obtained by simply appending an open-ended question, asking "why?" after the latter question. Since fisheries and coastal management projects include the potential for alternative occupations, it is important to obtain a bit more detail than that single question would generate; hence, several more questions are contained in the operational definition of the variable used here.

1. If you had your life to live over, would you still become a (fisher, fish farmer, and so on)?
   Why or why not?
2. If you could make the same income in an occupation other than (fisher, fish farmer, and so on), would you change your job?
   Why or why not?
3. What do you like about (fishing, fish farming, and so on) in comparison with other jobs you could do?

1.5.7 Ecological knowledge

It is possible to identify a folk taxonomic subgrouping of fish that is relatively highly differentiated in the project area and ask the respondent to list the names of the members of the subgroup. The number of terms recalled would indicate ecological knowledge. A fisher could also be requested to respond to a question about perceived trends in resource availability and reasons for the trends.

1. Is the (resource: amount of fish, crabs in the mangroves, whatever the fisheries and coastal management project focuses on) better off, worse off, or about the same as it has been over the past 10 years?
   If worse/better off, is it a little worse/better off, just worse/better off, or a lot worse/better off?
   Why?
2. What different types of resource (appropriate, highly differentiated types) do you know?

1.5.8 Occupational multiplicity

Occupations are a very important aspect of social structure, as well as an indicator of the relative importance of different aspects of the coastal resource. Secondary data is an inadequate source of information concerning occupations since most published statistics include only the full-time or primary occupation. Most coastal communities, especially in rural areas, are characterized by occupational multiplicity — a given individual or household may practice two, three, four or more income- or subsistence-producing activities. The only way to determine the distribution and relative importance of these activities is with a sample survey.

Ideally, one should obtain the value of all coastal activities that contribute to the household, such as the income earned from fishing and the value of fish brought home for food. The problem is that most primary producers in developing economies do not keep records of income, and income from fishing, for example, varies so much from day to day that it is difficult to accurately provide a figure for weekly or monthly income. It varies not only from day to day but from season to season. The difficulty of estimating income is compounded by the occupational multiplicity mentioned above.

If it is possible to obtain income values for these productive activities, do it. Experience, however, has indicated that an excessive amount of time is required to obtain this information, which usually turns out to be relatively unreliable. Since it is the relative importance of the activities that is significant to coastal planning, the relative importance of the activity to the individual household is the minimally acceptable level of measurement. This means that it is sufficient to obtain a ranking of the activities for each household. Questions that can be posed include the following. (An example of an analysis of occupational multiplicity follows the questions.)

1. Does your household have any sources of income other than (fishing, fish farming, other coastal resource–related principal occupation)? Does your household receive money from anyone living outside the household (for example, in the city, abroad?
   - Which of these activities or sources (for example, remittances from overseas would not be an activity) brings in the most income? Second most? Third most? and so on.
   - Do any of the activities provide over half your income?
2. Does your household have any sources of food other than what you buy or bring home from fishing/fish farming/other coast-related principal occupation?
   - In terms of food, which of these activities is the most important? Second-most important? Third-most important?
   - Do any of these activities provide more than half your food?

1.5.9 Assets

This is a substitute for a measure of income, which is frequently difficult to calculate in a reliable manner. It involves the following indicators, which must be further defined to be locally appropriate:

1. ownership of productive equipment (boat, pond, cages, as appropriate): owner, co-owner, group ownership (describe this), none
2. house structure: minimal, low, medium, high
3. furnishings: minimal, low, medium, high.

Box A1.4 Example of distribution of productive activities in a fishing village in North Sulawesi.

Most rural coastal villages manifest a great deal of occupational multiplicity. To investigate this, a sample survey was conducted on a random sample of 41 households in Bentenan, North Sulawesi, Indonesia (Pollnac et al. 1997). Respondents were asked to indicate all activities that contribute to household income and food, then to rank the activities. The table below presents percent distribution of ranking of productive activities for Bentenan, indicating a great deal of household occupational multiplicity.

Percent distribution of ranking of productive activities in Bentenan

Activity	1st	2nd	3rd	4th	5th	6th	Total
Fishing	54	05	02	05	—	—	66
Fry collecting	07	41	12	15	—	—	75
Seaweed farm	—	07	15	—	05	—	27
Fish trading	12	15	15	—	—	—	42
Fry trading	—	—	05	—	—	—	05
Seaweed trading	—	—	—	—	02	—	02
Other trading	02	05	02	02	—	—	11
Processing	—	05	10	20	—	—	35
Farming	12	12	17	07	02	02	52
Carpenter work	05	05	—	—	—	—	10
Boat building	—	05	—	—	—	—	05
Ornamental fish*	—	—	—	—	02	—	02
Resort work	—	—	—	05	—	—	05
Tailoring	02	—	—	—	—	—	02
Teaching	02	—	—	—	—	—	02
Raising animals	—	—	02	—	—	—	02
Remittance	02	—	—	—	—	—	02
Total	100**	100	80	54	11	02	

*capture of ornamental fish N=41

**column does not sum to 100 due to rounding

Glossary
Technical Terms, Abbreviations, and Acronyms

Adaptive management: Often applied to systems on which there is insufficient information; relies on feedback learning or learning-by-doing. Typically, experiments are designed to accelerate learning; policies may be used as experiments; and the distinction between the scientist, the manager, and the resource user are broken down.

Bioeconomic model: An analytical tool to facilitate management decisions. Bioeconomic models establish functional relationships between specific biological characteristics of the fishery resource and the economics of making use of the resource.

Bycatch: Part of a catch of a fishing unit taken incidentally in addition to the target species. Some or all of it may be returned to the sea as discards.

Capacity building: The sum of efforts needed to nurture, enhance, and utilize the skills and capabilities of people and institutions at all levels toward a particular goal, such as sustainable development.

Catch per unit effort (CPUE): The amount of catch that is taken per unit of fishing gear.

Census (fisheries): A survey in which the value of each variable for the survey area is obtained from the values of the variable in all reporting units.

Comanagement: A partnership arrangement in which government, the community of local resource users (fishers), external agents (non-governmental organizations, academic, and research institutions), and other fisheries and coastal resource stakeholders (boat owners, fish traders, money lenders, tourism establishments, etc.) share the responsibility and authority for decision-making in the management of a fishery.

Common property (common pool) resources: A class of resources for which exclusion (or control of access) is difficult, and where each user has the potential of subtracting from the welfare of all other users.

Community: A social group possessing shared beliefs and values, stable membership, and the expectation of continued interaction. It can be bounded geographically, by political or resource boundaries, or socially as a community of individuals with common interests.

Community-based resource management (CBRM): A central element of comanagement. CBRM is people-centred and community-focused, having a narrower scope and scale than comanagement. Government most often plays a minor role in CBRM, providing mainly legitimacy and accountability, since only government can legally establish and defend user rights and security of tenure at the community level.

Community-centred comanagement (CCCM): Includes the characteristics of both CBRM and comanagement; that is, is people-centred, community-oriented, resource-based, and partnership-based. It focuses on the community but recognizes that to sustain such action, horizontal and vertical links are necessary and meaningful partnerships can occur only when the community is empowered and organized.

Dataless management: This term describes a fishery management approach prescribed by Johannes in which management proceeds using available information from a variety of sources and is not delayed due to lack of technical information.

Divisibility: The feasibility or extent to which a common property resource can be divided up for private possession; the question of boundary conditions that applies to the management of a resource such as a fish stock.

Ecological resilience: A measure of flexibility of an ecosystem to maintain its structure and function. The ability of an ecosystem to absorb change and still persist.

Ecosystem-based management: Resource management that takes account of interactions of a given resource with other components in the ecosystem in which it is a part.

Empowerment: Having the power and responsibility to do something; the ability of a person or a group of people to control or to have an input into decisions that affect their livelihoods.

Exclusion problem: The problem of how to control access to a resource, given that it is difficult or costly to exclude potential users from gaining access.

Exclusive Economic Zone (EEZ): All waters beyond and adjacent to the territorial sea up to a maximum of 200 nautical miles (including territorial sea). In the EEZ, the state has sovereign rights and responsibilities as defined in the *UN Convention on the Law of the Sea* (UNCLOS).

Fisher: A person (male or female) participating in a fishery (in preference to the previously used term "fisherman"). An individual who takes part in fishing conducted from a fishing vessel, platform (whether fixed or floating) or from the shore.

Fisheries Management Plan (FMP): A plan to achieve specified management goals and objectives for a fishery or set of fisheries. It includes data collection, analyses, and management measures for the fishery.

Fishery Ecosystem Plan (FEP): A plan that addresses the problems and needs of fisheries at the ecosystem level. This differs from the usual management plan that deals specifically with the exploited resource. In the USA, an FEP is required under the *Sustainable Fisheries Act*.

Fishing effort: The amount of time or fishing power used to harvest fish. Fishing power can be expressed in terms of gear size and quantity, boat size, horsepower, fuel consumption, manpower, etc.

Fishing mortality (FM): A mathematical expression of the rate of deaths of fish due to fishing.

Folk taxonomy: Local names for fish, plants, boats, etc., hierarchically organized by the process of inclusion. Contrasts with the "common name" due to its focus on the ordering of the terms.

Geographic Information System (GIS): An information system that stores and manipulates data that is referenced to locations on the earth's surface, such as digital maps and sample locations.

Household: A basic unit for sociocultural and economic analysis. It includes all persons, kin and nonkin, who live in the same dwelling and share income, expenses and daily subsistence tasks. The concept of household is based on the arrangements made by persons, individually or in groups, for providing themselves with food or other essentials for living.

Index of abundance: A relative measure of the abundance of a stock; for example, a time series of catch per unit of effort data.

Indicator: A variable, pointer, or index. Its fluctuation reveals the variations in key elements of a system. The position and trend of the indicator in relation to reference points or values indicate the present state and dynamics of the system. Indicators provide a bridge between objectives and action.

Indigenous knowledge: Local knowledge held by a group of indigenous people, or local knowledge unique to a given culture or society. Traditional ecological knowledge is a subset of indigenous knowledge.

Individual transferable quota (ITQ): A quantitative harvesting right that can be traded in the open market.

Information management: Managing a structured set of processes, people, and equipment for converting data into information and then using it for specified purposes.

Institutions: Socially constructed codes of conduct that define practices, assign roles, and guide interactions; the set of rules actually used.

Large Marine Ecosystem (LME): Relatively large regions of the oceans, on the order of 200 000 km^2, or larger, characterized by distinct bathymetry, hydrography, productivity, and trophically dependent populations, within which the interdependencies are sufficiently strong to warrant a holistic management approach.

Limit Reference Point (LRP): Indicates the limit beyond which the state of a fishery and/or a resource is not considered desirable.

Local knowledge: Knowledge based on local observations made by resource users; differs from traditional knowledge in not being multigenerational or culturally transmitted.

Management authority: The legal entity that has been assigned by a state or states with a mandate to perform certain specified management functions in relation to a fishery, or an area (e.g. a coastal zone). Generally used to refer to a state authority, the term may also refer to a local or international management organization.

Management objective driven (MOD): An approach to fishery management in which research, assessment, and management measures are based primarily upon the desired management objectives.

Management objective: A formally established state of the fishery that is actively sought and provides a direction for management action.

Management reference direction (MRD): A direction in which management seeks to take a fishery through action in a case where there is insufficient information or resources to specify an exact target.

Management unit: A fishery unit, including the resource and the fishers that is known, or assumed to be, sufficiently discrete that it may be managed separately from other units and cannot be effectively managed on a smaller scale.

Marine protected area (MPA): A spatially defined area in which all populations are free of exploitation.

Maximum economic yield (MEY): The level of overall yield from a fishery that provides the maximum economic return as defined by the difference between the monetary cost of fishing and the monetary value of the yield.

Maximum sustainable yield (MSY): The largest average catch that can be taken continuously (sustained) from a stock under average environmental conditions. MSY is often used synonymously with the term Potential Yield as the target reference point to guide fisheries managers in resource utilization.

Monitoring: The collection of information for the purpose of assessing progress and impacts.

Natural mortality (M): Deaths of fish from all natural causes except fishing.

Occupational pluralism or multiplicity: The situation where a person derives their income from several types of work done in parallel throughout the year, or sequentially (seasonally).

Open access: Free-for-all; resources freely open to any user; absence of property rights

Optimum sustainable yield (OSY): A level of yield that is defined based on a combination and rationalization all of the outputs that are considered to be important for the fishery in question, provided that these outputs are sustainable.

Policy: The course of action for an undertaking adopted by a government, a person or another party.

Precautionary approach: A set of measures taken to implement the precautionary principle. That is, a set of agreed cost-effective measures and actions, including future courses of action, which ensures prudent foresight and reduces or avoids risk to the resource, the environment, and the people, to the extent possible, taking explicitly into account existing uncertainties and the potential consequences of being wrong.

Property rights: Claim to a benefit stream that is collectively protected, in most cases by the state.

Quota: A share of the Total Allowable Catch (TAC) allocated to an operating unit, such as a country, a vessel, a company, or an individual fisherman (individual quota), depending on the system of allocation.

Reference point: An estimated value derived from an agreed scientific procedure and/or model, which corresponds to a specific state of the resource and of the fishery, and that can be used as a guide for fisheries management. A reference point indicates a particular state of a fishery indicator corresponding to a situation considered as desirable (Target Reference Point) or undesirable and requiring immediate action (Limit Reference Point).

Resilience: See Ecological resilience

Shared stocks: Fish stocks that occur at some point in their life history in the waters of more than one country and hence are shared by the fishers of those countries. Responsibility for management must also be shared. Stocks may also be shared between jurisdictions within countries.

Social capital: Features of social organization, such as trust, norms, and networks. A group with a high degree of trust among its members, shared values, and extensive networks to share information or resources is said to have high social capital.

Social-ecological system: A term used to emphasize the point that social and ecological systems are in fact linked, and that the delineation between social and ecological (and between nature and culture) is artificial and arbitrary; the integrated concept of humans-in-nature.

Stakeholders: Individuals or groups (including governmental and non-governmental institutions, traditional communities, universities, research institutions, development agencies and banks, and donors) with an interest or claim.

Stakeholder analysis: A process that seeks to identify and describe the interests of all the stakeholders in a fishery. It is considered to be a necessary precursor to participatory management.

Stock assessment: The process of collecting and analyzing biological and statistical information to determine the changes in the abundance of fishery stocks in response to fishing and, to the extent possible, to predict future trends of stock abundance.

Stock assessment driven (SAD): An approach to fishery management in which conventional quantitative stock assessment, aimed at estimating present and desired levels of fishing mortality, is considered to be a prerequisite to management and becomes the top priority activity.

Stock: A grouping of fish usually based on genetic relationship, geographic distribution and movement patterns that can be considered a discrete entity for management purposes.

Straddling stock: Stock that occurs both within the EEZ and in an area beyond and adjacent to EEZ [Article 63(2) of UNCLOS].

Subtractability: How each person's use of the resource subtracts from the welfare of the others.

Target Reference Point (TRP): Corresponds to a state of a fishery and/or a resource that is considered desirable.

Target species: Those species that are primarily sought by the fishermen in a particular fishery.

Total Allowable Catch (TAC): Total catch allowed to be taken from a resource in a specified period (usually a year), as defined in the management plan.

Total mortality rate (Z): The combined effect of all sources of mortality acting on a fish population.

Traditional ecological knowledge: A cumulative body of knowledge, practice, and belief, evolving by adaptive processes and handed down through generations by cultural transmission, about the relationship of living beings (including humans) with one another and with their environment.

Traditional knowledge: A cumulative body of knowledge, practice and belief evolving by adaptive processes and handed down through generations by cultural transmission.

Tragedy of the commons: A metaphor formulated by Garrett Hardin to explain the individually rational use of a resource held in common in a way that eventually brings ruin to all who depend on the resource.

UNCLOS: *United Nations Convention on the Law of the Sea.*

Utility maximizing: The maximization of satisfaction an individual gets from the services provided by the commodity or commodities consumed during a given period.

Variable: A quantity that varies or may vary. Part of a mathematical expression or model that may assume any value, sometimes within specified limits.

Vessel Monitoring System (VMS): VMS provides monitoring agencies with accurate locations of fishing vessels that are participating in the VMS.

Yield: Catch in weight. Catch and yield are often used interchangeably.

The major source for this glossary was FAO 1999 *Guidelines for the routine collection of capture fishery data*. Prepared at the FAO/DANIDA Expert Consultation, Bangkok, Thailand, 18-30 May 1998. FAO Fisheries Technical Paper No. 382. FAO, Rome; another important source was Barbados Fisheries Management Plan 1997–2000

About the Authors

Fikret Berkes is Professor of Natural Resources at the University of Manitoba, Winnipeg, Canada. He holds a PhD degree in Marine Sciences from McGill University, Montreal. His main area of interest is common property resources and community-based resource management. He has devoted most of his professional life to investigating the interrelations between societies and their resources, and to examining the conditions under which the "tragedy of the commons" may be avoided. He has experience with small-scale fisheries in the Canadian North, the Caribbean, Turkey, South Asia, and East Africa. His recent publications include two books, *Sacred Ecology* (Taylor & Francis, 1999) and *Linking Social and Ecological Systems* (with Carl Folke, Cambridge University Press, 1998).

Robin Mahon is a Barbadian fisheries consultant. Previously, he worked for the Canada Department of Fisheries and Oceans, FAO, and the Caribbean Community (CARICOM). He holds a PhD degree in aquatic sciences from the University of Guelph, Canada. His areas of interests include all aspects of the sustainable use of aquatic living resources. He has spent most of his professional life on assessment and management of small-scale fisheries and on the integration of developing countries into the international fisheries regime. He has progressively been involved in marine resource assessment, management, policy development, and institutional strengthening at national and regional levels. He is committed to promoting the conservation of the Caribbean Sea and to its recognition by all Caribbean peoples as a common, integrating resource.

Patrick McConney is Chief Fisheries Officer in the Fisheries Division of the Ministry of Agriculture in Barbados. He joined the government as a Marine Biologist, later completing an interdisciplinary Masters degree in environmental studies (Dalhousie University) and an interdisciplinary PhD in resource management (University of British Columbia). His responsibilities cover all aspects of fisheries management, conservation and development, including aquaculture. Recent research interests include participatory fisheries planning and management, systems of governance, and small island developing state (SIDS) issues such as approaches to fisheries science and technology appropriate to small-scale fisheries.

Richard B. Pollnac is Professor of Anthropology at the University of Rhode Island. He has conducted research and worked on fishery and coastal management projects in New England (USA), Latin America, the Caribbean, East and West Africa, the Middle East, the South Pacific, and South East Asia over the past quarter century. His most recent publications are two books, *Assessing Behavioural Aspects of Coastal Resource Use* (with Brian R. Crawford) and *Rapid Assessment of Management Parameters for Coral Reefs* (2000 and 1998 respectively) and the article "Unexpected relationships between coral reef health and socioeconomic pressures in the Philippines: ReefBase/RAMP applied" (with J.W. McManus, A.E. delRosario, A.A. Banzon, S.G. Vergara and M.L.G. Gorospe), which appeared in *Marine and Freshwater Research* 51, 529–533, 2000.

Robert S. Pomeroy is Senior Associate, Coastal and Marine Projects, in the Biological Resources Program of the World Resources Institute (WRI) in Washington, DC. He holds a PhD degree in Natural Resource Economics from Cornell University. His main areas of interest are policy analysis, fisheries management and development, aquaculture economics, and coastal area management. Over the past 15 years, he has worked on fisheries research and development projects in over 40 countries in Asia, Africa, Latin America, and the Caribbean. Before joining WRI in 1999, he was a Senior Scientist at the International Center for Living Aquatic Resources Management (ICLARM) in Manila, Philippines, where among other projects, he was the leader of the fisheries comanagement research project.

References

Abdullah, A.; Hamad, A.S.; Ali, A.M.; Wild, R.G. 2000. Misali Island, Tanzania: an open-access resource redefined. Papers of the IASCP (International Association for the Study of Common Property), June 2000, Bloomington, IN, USA. http://www.indiana.edu/~iascp2000.htm.

Acheson, J. 1988. The lobster gangs of Maine. University Press of New England, Hanover, NH, USA.

Agbayani, R.F.; Siar, S.V. 1994. Problems encountered in the implementation of a community-based fishery resource management project. *In* Pomeroy, R.S., ed., Community management and common property of coastal fisheries in Asia and the Pacific: concepts, methods and experiences. ICLARM (International Center for Living Aquatic Resources Management) Conference Proceedings 45. ICLARM, Manila, Philippines. pp. 115–123.

Ahmed, M.; Capistrano, A.D.; Hossain, M. 1997. Experience of partnership models for the co-management of Bangladesh fisheries. Fisheries Management and Ecology, 4, 233–248.

Aiken, K.A.; Kong, G.A.; Smikle, S.; Mahon, R.; Appeldoorn, R. 1999. The queen conch fishery on Pedro Bank, Jamaica: discovery, development, management. Ocean and Coastal Management, 42, 1069–1081.

Akimichi, T. 1984. Territorial regulation in the small-scale fisheries of Itoman, Okinawa. *In* Ruddle, K.; Akimichi, T., ed., Maritime institutions in the Western Pacific. Senri Ethnological Studies 17, National Museum of Ethnology, Osaka, Japan. pp. 89–120.

Alcala, A.C.; Vande Vusse, F.J. 1994. The role of government in coastal resources management. *In* Pomeroy, R.S., ed., Community management and common property of coastal fisheries in Asia and the Pacific: concepts, methods and experiences. ICLARM (International Center for Living Aquatic Resources Management) Conference Proceedings 45. ICLARM, Manila, Philippines, pp. 12–19.

Alverson, D.L.; Freeberg, M.H.; Murawski, S.A.; Pope, J.G. 1994. A global assessment of fisheries bycatch and discards. Food and Agriculture Organization, Rome, Italy. Fisheries Technical Paper No. 339.

Amarasinghe, U.S.; Chandrasekara, W.U.; Kithsiri, H.M.P. 1997. Traditional practices for resource sharing in an artisanal fishery of a Sri Lankan estuary. Asian Fisheries Science, 9, 311–323.

Anonuevo, C.T. 1994. The role of non-governmental organizations in community-based coastal resources management. *In* Pomeroy, R.S., ed., Community management and common property of coastal fisheries in Asia and the Pacific: concepts, methods and experiences. ICLARM (International Center for Living Aquatic Resources Management) Conference Proceedings 45. ICLARM, Manila, Philippines, 145–148.

Appeldoorn, R.S. 1996. Model and method in reef fishery assessment. *In* Polunin, N.V.C.; Roberts, C.M., ed., Reef fisheries. Chapman and Hall, New York, NY, USA. pp. 219–248.

______ 1998. Ecological goals for marine fishery reserve design: workshop summary. Proceedings of the Gulf Caribbean Fish Institute, 50, 294–303.

Arnason, R.; Hannesson, R.; Schrank, W.E. 2000. Costs of fisheries management: the cases of Iceland, Norway and Newfoundland. Marine Policy, 24, 233–243.

Atapattu, A.R. 1987. Territorial use rights in fisheries in Sri Lanka. Symposium on the exploitation and management of marine fishery resources in Southeast Asia. Indo-Pacific Fisheries Commission. RAPA Report 1987 (10), 379–401.

Axelrod, R. 1984. The evolution of cooperation. Basic Books, New York, NY, USA.

Baines, G.B.K. 1989. Traditional resource management in the Melanesian South Pacific: a development dilemma. *In* Berkes, F., ed., Common property resources. Belhaven, London, UK. pp. 273–295.

Baines, G.; Hviding, E. 1992. Traditional environmental knowledge from the Marovo area of the Solomon Islands. *In* Johnson, M., ed., Lore. International Development Research Centre, Ottawa, ON, Canada.

Baland, J.-M.; Platteau, J.-P. 1996. Halting degradation of natural resources. Is there a role for rural communities? Food and Agriculture Organization/Clarendon Press, Oxford, UK.

Beddington, J.R.; Rettig, R.B. 1983. Approaches to the regulation of fishing effort. Food and Agriculture Organization, Rome, Italy. Fisheries Technical Paper No. 243.

Berkes, F. 1985. Fishermen and the "tragedy of the commons." Environmental Conservation, 12, 199–205.

______ 1986. Local-level management and the commons problem: a comparative study of Turkish coastal fisheries. Marine Policy, 10, 215–229.

______ 1987. The common property resource problem and the fisheries of Barbados and Jamaica. Environmental Management, 11, 225–235.

Berkes, F., ed. 1989. Common property resources: ecology and community-based sustainable development. Belhaven, London, UK.

Berkes, F. 1992. Success and failure in marine coastal fisheries of Turkey. *In* Bromley, D.W., ed., Making the Commons Work. Institute for Contemporary Studies Press, San Francisco, CA, USA. pp. 161–182.

______ 1994a. Co-management: bridging the two solitudes. Northern Perspectives, 22(2-3), 18–20.

______ 1994b. Property rights and coastal fisheries. *In* Pomeroy, R.S., ed., Community management and common property of coastal fisheries in Asia and the Pacific: concepts, methods and experiences. ICLARM (International Center for Living Aquatic Resources Management) Conference Proceedings 45. ICLARM, Manila, Philippines. pp. 51–62.

______ 1999. Sacred ecology. Traditional ecological knowledge and resource management. Taylor & Francis, Philadelphia, PA, USA and London, UK.

Berkes, F.; Folke, C., ed. 1998. Linking social and ecological systems. Management practices and social mechanisms for building resilience. Cambridge University Press, Cambridge, UK.

Berkes, F.; Folke, C.; Gadgil, M. 1995. Traditional ecological knowledge, biodiversity, resilience and sustainability. *In* Perrings, C.A.; Mäler, K.G.; Folke, C.; Jansson, B.O.; Holling, C.S., ed., Biodiversity conservation. Kluwer Academic Publishers, Dordrecht, Netherlands. pp. 281–299.

Berkes, F.; George, P.; Preston, R.J. Co-management: the evolution in theory and practice of the joint administration of living resources. Alternatives, 18(2), pp. 12–18.

Berkes, F.; Kislalioglu, M. 1989. A comparative study of yield, investment and energy use in small-scale fisheries. Fisheries Research, 7, 207–224.

Berkes, F.; Kislalioglu, M.; Folke, C.; Gadgil, M. 1998. Exploring the basic ecological unit: ecosystem-like concepts in traditional societies. Ecosystems, 1, 409–415.

Berlin, B. 1992. Ethnobiological classification: principles of categorization of plants and animals in traditional societies. Princeton University Press, Princeton, NJ, USA.

Bernard, H.R.; Killworth, P.; Kronenfeld, D.; Sailer, L. 1984. The problem of informant accuracy: the validity of retrospective data. Annual Reviews in Anthropology, 1984, 495–517.

Blaxter, J.H.S. 2000. The enhancement of marine fish stocks. Advances in Marine Biology, 38, 1–58.

Borrini-Feyerabend, G. 1996. Collaborative management of protected areas: tailoring the approach to the context. IUCN (The World Conservation Union), Gland, Switzerland.

Borrini-Feyerabend, G., compiler. 1997. Beyond fences: seeking social sustainability in conservation. IUCN (The World Conservation Union), Gland, Switzerland.

Bromley, D.W., ed. 1992. Making the commons work. Institute of Contemporary Studies Press, San Francisco, CA, USA.

Bromley, D.W.; Cernea, M.M. 1989. The management of common property natural resources: some conceptual and operational fallacies. World Bank, Washington, DC, USA. World Bank Discussion Paper 57,

Brown, D.N.; Pomeroy, R.S. 1999. Co-management of CARICOM (Caribbean Community) fisheries. Marine Policy, 23, 549–570.

Buckworth, R.C. 1998. World fisheries are in crisis? We must respond! *In* Pitcher, T.J.; Hart, P.J.B.; Pauly, D., ed., Reinventing fisheries management. Kluwer Academic Publishers, London, UK. pp. 3–17.

Burbidge, J. 1997. Beyond prince and merchant. Pact Publications, New York, NY, USA.

Bush, R.A.B.; Folger, J.P. 1994. The promise of mediation: responding to conflict through empowerment and recognition. Jossey-Bass, San Francisco, CA, USA.

Butler, J.N.; Burnett-Herkes, J.; Barnes, J.A.; Ward, J. 1993. The Bermuda fisheries — a tragedy of the commons averted? Environment, 35(1), 7–33.

Butler, M.J.A.; LeBlanc, C.; Belbin, J.A.; MacNeill, J.L. 1986. Marine resource mapping: an introductory manual. Food and Agriculture Organization, Rome, Italy. Fisheries Technical Paper No. 274.

Caddy, J.F. 1998. A short review of precautionary reference points and some proposal for their use in data-poor situations. FAO Fisheries Technical Paper No. 379, 29 pp.

______ 1999. Fisheries management in the twenty-first century: will new paradigms apply? Reviews in Fish Biology, 9, 1–43.

Caddy, J.F.; Bazigos, G. P. 1985. Guidelines for statistical monitoring of fisheries in man-power limited situations. Food and Agriculture Organization, Rome, Italy. Fisheries Technical Paper No. 257,

Caddy, J.F.; Mahon, R. 1995. Fishery management reference points. Food and Agriculture Organization, Rome, Italy. Fisheries Technical Paper No. 347, 87 pp.

Caddy, J.F.; Sharp, G.D. 1986. An ecological framework for marine fishery investigations. Food and Agriculture Organization, Rome, Italy. Fisheries Technical Paper No. 283,

CCAMF (Caribbean Coastal Area Management Foundation). 1998. Portland Bight Sustainable Development Area, Jamaica: management plan, 1998–2003. CCAMF, Kingston, Jamaica.

Cernea, M.M. 1987. Farmer organizations and institution building for sustainable development. Regional Development Dialogue, 8(2), 2–7.

Cernea, M.M., ed. 1991. Putting people first: sociological variables in rural development (2nd ed.). Oxford University Press, New York, NY, USA.

Chakalall, B.; Mahon, R.; McConney, P. 1998. Current issues in fisheries governance in the CARICOM (Caribbean community). Marine Policy, 22, 29–44.

Chambers, R. 1983. Rural development: putting the last first. Longman Scientific & Technical, Essex, UK.

______ 1997. Whose reality counts? Putting the first last. Intermediate Technology Publications, London, UK.

Chapman, M.D. 1987. Traditional political structure and conservation in Oceania. Ambio, 16, 201–205.

Charles, A.T. 1998a. Living with uncertainty in fisheries: analytical methods, management priorities and the Canadian groundfishery experience. Fisheries Research, 37, 37–50.

______ 1998b. Beyond the status quo: rethinking fishery management. *In* Pitcher, T.J.; Hart, P.J.B.; Pauly, D., ed., Reinventing fisheries management. Kluwer, London, UK. pp. 111–112.

______ 2001. Sustainable fishery systems. Blackwell Science, Oxford, UK.

Christensen, M.S.; De Melo Soares, W.J.; Bezerra E.; Silva, F.C.; Leite Barros, G.M. 1995. Participatory management of a reservoir fishery in northeastern Brazil. Naga, 18, 7–9.

Christensen, V; Pauly, D. 1993. Trophic models of aquatic ecosystems. International Center for Living Aquatic Resources Management Conference Proceedings No. 29. International Center for Living Aquatic Resources Management, Manila, Philippines.

Christie, P.; White, A.T. 1997. Trends in development of coastal area management in tropical countries: from central to community orientation. Coastal Management, 25, 155–181.

Christy, F.T., Jr. 1982. Territorial use rights in marine fisheries: definitions and conditions. Food and Agriculture Organization, Rome, Italy. Fisheries Technical Paper No. 227,

Christy, F.T., Jr.; Scott, A.D. 1965. The common wealth in ocean fisheries. Johns Hopkins University Press, Baltimore, MD, USA.

Ciriacy-Wantrup, S.V.; Bishop, R.C. 1975. "Common property" as a concept in natural resources policy. Natural Resources Journal, 15, 713–727.

Clark, C.W. 1973. The economics of overexploitation. Science, 181, 630–634.

______ 1976. Mathematical bioeconomics. Wiley, New York, NY, USA.

______ 1985. Bioeconomic modelling and fisheries management. Wiley, New York, NY, USA.

Cochrane, K.L.; Butterworth, D.S.; De Oliveira, J.A.A.; Roel, B.A. 1998. Management procedures in a fishery based on highly variable stocks and with conflicting objectives: experiences in the South African pelagic fishery. Reviews in Fish Biology and Fisheries, 8, 177–214.

Cohen, J. 1988. Statistical power analysis for the behavioural sciences. Lawrence Erlbaum Associates, Hillsdale, NJ, USA.

Commission of the European Communities. 1993. Project cycle management, integrated approach and logical framework. Commission of the European Communities, Brussels, Belgium.

Copes, P. 1986. A critical review of the individual quota as a device in fisheries management. Land Economics, 62, 278–291.

Cordell, J., ed. 1989. A sea of small boats. Cultural Survival Inc., Cambridge, MA, USA.

Corkett, C.J. 1997. Managing the fisheries by social engineering: a re-evaluation of the methods of stock assessment. Journal of Applied Ichthyology, 13, 159–170.

Costa-Pierce, B.A. 1987. Aquaculture in ancient Hawaii. BioScience 37, 320–330.

______ 1988. Traditional fisheries and dualism in Indonesia. Naga, 11(2), 34 pp.

Dalzell, P.; Adams, T.; Polunin, N. 1996. Coastal fisheries in the Pacific Islands: an annual review. Oceanography and Marine Biology, 34, 395–531.

Davis, J.; Whittington, D. 1998. "Participatory" research for development projects: a comparison of the community meeting and household survey techniques. Economic Development and Cultural Change, 47(1), 73–94.

Dayton, P.K.; Thrush, S.F.; Agardy, M.T.; Hofman, R.J. 1995. Environmental effects of marine fishing. Aquatic conservation: marine and freshwater ecosystem, 5, 205–232.

Deere, C. 1999. Eco-labeling and sustainable fisheries. IUCN (The World Conservation Union), Washington, DC, USA.

______ 2000. Net gains: linking fisheries management, international trade and sustainable development. IUCN (The World Conservation Union), Washington, DC, USA.

Derman, B.; Ferguson, A. 1995. Human rights, environment, and development: the dispossession of fishing communities on Lake Malawi. Human Ecology, 23, 125–142.

Diaz-de-Leon, A.; Seijo, J.C. 1992. A multi-criteria non-linear optimization model for the control and management of a tropical fishery. Marine Resource Economics, 7, 23–40.

Domeier, M.L.; Colin, P.L. 1997. Tropical reef fish spawning aggregations: defined and reviewed. Bulletin of Marine Science, 60, 698–726.

Doulman, D.J. 1993. Community-based fishery management: toward the restoration of traditional practices in the South Pacific. Marine Policy, 17(2), 108–117.

Dyer, C.L.; McGoodwin, J.R., ed. 1994. Folk management in the world's fisheries. University Press of Colorado, Niwot, CO, USA.

English, S.; Wilkinson, C.; Baker, V., ed. 1997. Survey manual for tropical marine resources (2nd ed.). Australian Institute of Marine Science, Townsville, Australia.

EPAP (Ecosystem Principles Advisory Panel). 1999. Ecosystem-based fishery management. A report to Congress by the EPAP. National Oceanic and Atmospheric Administration, National Marine Fisheries Service, Silver Spring, MD, USA.

FAO (Food and Agriculture Organization). 1993. Marine resources of the Antilles. FAO, Rome, Italy. FAO Fisheries Technical Paper No. 326,

______ 1995. Code of conduct for responsible fisheries. FAO, Rome, Italy.

______ 1996a. Integration of fisheries into coastal area management. Fisheries Department of FAO, Rome, Italy.

______ 1996b. Precautionary approach to capture fisheries and species introductions. Elaborated by the Technical Consultation on the Precautionary Approach to Capture Fisheries (Including Species Introductions). FAO, Rome, Italy. FAO Technical Guidelines for Responsible Fisheries No. 2,

______ 1997. Fisheries management. FAO, Rome, Italy. FAO Technical Guidelines for Responsible Fisheries No. 4,

______ 1998. Responsible fish utilization. FAO, Rome, Italy. FAO Technical Guidelines for Responsible Fisheries No. 7,

______ 1999. Guidelines for the routine collection of capture fishery data. FAO, Rome, Italy. FAO Fisheries Technical Paper No. 382

______ Fishery Resources Division. 1999. Indicators for sustainable development of marine capture fisheries. FAO, Rome, Italy. FAO Technical Guidelines for Responsible Fisheries No. 8,

Feeny, D. 1994. Frameworks for understanding resource management on the commons. *In* Pomeroy, R.S., ed. Community management and common property of coastal fisheries in Asia and the Pacific: concepts, methods and experiences. ICLARM (International Center for Living Aquatic Resources Management) Conference Proceedings 45. ICLARM, Manila, Philippines.

______ 1998. Suboptimality and transaction costs on the commons. *In* Loehman, E.T.; Kilgour, D.M., ed., Designing institutions for environment and resource management. Elgar, Cheltenham, UK. pp. 124–141.

Feeny, D.; Berkes, F.; McCay, B.J.; Acheson, J.M. 1990. The tragedy of the commons:

Twenty-two years later. Human Ecology, 18, 1–19.
Felt, L.F. 1990. Barriers to user participation in the management of the Canadian Atlantic salmon fishery: if wishes were fishes. Marine Policy, 14(4), 345–360.
______ 1994. Two tales of a fish: the social construction of indigenous knowledge among Atlantic Canadian salmon fishers. *In* Dyer, C.L.; McGoodwin, J.R., ed., Folk management in the world's fisheries. University Press of Colorado, Niwot, CO, USA. pp. 251–286.
Fisher, R.; Ertel, D. 1995. Getting ready to negotiate: a step-by-step guide to preparing for any negotiation. Penguin Books, New York, NY, USA.
Fisher, R.; Ury, W; Patton, B. 1991. Getting to yes. Penguin Books, New York, NY, USA.
Fisheries Division. 1997. Barbados fisheries management plan. Ministry of Agriculture and Rural Development, Government of Barbados, Bridgetown, Barbados.
Fogarty, M. 1999. Essential habitat, marine reserves and fishery management. Trends in Ecology and Evolution, 14, 133–134.
Fraser, N.M.; Siahainenia, A.J.; Kasmedi, M. 1998. Preliminary results of participatory manta tow training: Blongko, North Sulawesi. Jurnal Pengelolaan Sumberdaya Pesisir dan Lautan Indonesia, 1(1), 31–35.
Freeman, M.M.R.; Matsuda, Y.; Ruddle, K., ed. 1991. Adaptive marine resource management systems in the Pacific. Special issue of Resource Management and Optimization, 8(3/4), 127–245.

Gadgil, M.; Seshagiri Rao, P.R.; Utkarsh, G.; Pramod, P.; Chhatre, A. 2000. New meanings for old knowledge: the People's Biodiversity Registers Program. Ecological Applications, 10, 1251–1262.
Garcia, S.M. 1994. The precautionary approach: its implications in capture fisheries management. Ocean and Coastal Management, 22, 99–125.
Garcia, S.M.; Newton, C. 1994. Current situation, trends and prospects in world capture fisheries. Paper presented at the Conference on Fisheries Management, Global Trends, June 1994, Seattle, Washington, DC, USA.
Giddens, A. 1994. Beyond left and right. The future of radical politics. Stanford University Press, Stanford, CA, USA.
Gomes, C.; Mahon, R.; Hunte, W.; Singh-Renton, S. 1998. The role of drifting objects in pelagic fisheries in the southeastern Caribbean. Fisheries Research, 34, 47–58.
Gordon, J.S. 1954. The economic theory of a common-property resource: the fishery. Journal of Political Economy, 62, 124–142.
Grenier, L. 1998. Working with indigenous knowledge. A guide for researchers. International Development Research Centre, Ottawa, ON, Canada.
Gulland, J.A. 1974. The management of marine fisheries. University of Washington Press, Seattle, Washington, DC, USA.
Gunderson, L.; Holling, C.S.; Light, S., ed. 1995. Barriers and bridges to the renewal of ecosystems and institutions. Columbia University Press, New York, NY, USA.

Guttierrez, J.S.; Rivera, R.A.; Dela Cruz, Q.L. 1995. SCAD (Sustainable coastal area development) Program: Barili, Cebu. *In* Foltz, C.; Pomeroy, R.S.; Barber, C.V., ed., Proceedings of the Visayas-wide conference on community-based coastal resources management and fisheries co-management. ICLARM (International Center for Living Aquatic Resources Management) Manila, Philippines. Fisheries Co-Management Project Working Paper 4.

Haggan, N. 1998. Reinventing the tree: reflections on the organic growth and creative pruning of fisheries management structures. *In* Pitcher, T.J.; Hart, P.J.B.; Pauly, D., ed., Reinventing fisheries management. Kluwer Academic Publishers, London, UK.

Hamlisch, R. 1988. Methodology and guidelines for fisheries development planning with special reference to developing countries in the African region. Food and Agriculture Organization, Rome, Italy. Fisheries Technical Paper No. 297,

Hanna, S.S. 1996. User participation and fishery management performance within the Pacific fishery management council. Ocean and Coastal Management, 28, 23–44.

Hardin, G. 1968. The tragedy of the commons. Science, 162, 1243–1248.

Harkes, I.; Novaczek, I. 2000. Institutional resilience of *sasi laut*, a fisheries management system in Indonesia. Papers of the IASCP (International Association for the Study of Common Property), June 2000, Bloomington, IN, USA. http://www.indiana.edu/~iascp2000.htm.

Healey, M.C. 1984. Multiattribute analysis and the concept of optimum yield. Canadian Journal of Fisheries and Aquatic Sciences, 41, 1393–1406.

Hedayat, A.; Sinha, B.K. 1991. Design and inference in finite population sampling. Wiley, New York, NY, USA.

Henry, G.T. 1990. Practical sampling. Sage Publications, London, UK.

Hilborn, R. 1992. Can fisheries agencies learn from experience? Fisheries, 17(4), 6–14.

Hilborn, R.; Ludwig, D. 1993. The limits of applied ecological research. Ecological Applications, 3, 550–552.

Hilborn, R.; Walters, C.J. 1992. Quantitative fisheries stock assessment. Chapman and Hall, London, UK.

Holling, C.S., ed. 1978. Adaptive environmental assessment and management. Wiley, London, UK.

Holling, C.S. 1986. The resilience of terrestrial ecosystems: local surprise and global change. *In* Clark, W.C.; Munn, R.E., ed., Sustainable development of the biosphere. Cambridge University Press, Cambridge, UK. pp. 292–317.

Holling, C.S.; Meffe, G.K. 1996. Command and control and the pathology of natural resource management. Conservation Biology, 10, 328–337.

Holling, C.S.; Schindler, D.W.; Walker, B.W.; Roughgarden, J. 1995. Biodiversity in the functioning of ecosystems: an ecological synthesis. *In* Perrings, C.; Mäler, K.-G.; Folke, C.; Holling, C.S.; Jansson, B.-O., ed., Biodiversity loss: economic and ecological issues. Cambridge University Press, Cambridge, UK.

Holman, P.; Devane, T., ed. 1999. The change handbook: group methods for shaping the future. Berrett-Koehler, San Francisco, CA, USA.

Hunt, C. 1997. Cooperative approaches to marine resources management in the South Pacific. *In* Larmour, P., ed., The governance of common property in the Pacific Region Australian National University, Canberra, Australia. pp.145–164.

Huntington, H.P. 1998. Observations on the utility of the semi-directive interview for documenting traditional ecological knowledge. Arctic, 51, 237–242.

Hviding, E.; Jul-Larsen, E. 1993. Community-based resource management in tropical fisheries. Centre for Development Studies, University of Bergen, Norway.

IADB (Inter-American Development Bank). 1997. Evaluation — a management tool for improving project performance. Logical Framework Analysis. IADB, Washington, DC, USA.

ICLARM (International Center for Living Aquatic Resources Management). 1999. ICLARM's Strategic Plan 2000–2020. Draft. ICLARM, Manila, Philippines.

IIRR (International Institute of Rural Reconstruction). 1998. Participatory methods in community-based coastal resources management, 3 volumes. IIHR, Silang, Cavite, Philippines.

Jasanoff, S.; Colwell, R.; Dresselhaus, M.S.; et al. 1997. Conversations with the community: AAAS (American Association for the Advancement of Science) at the millennium. Science, 278, 2066–2067.

Jennings, S.; Greenstreet, S.P.R.; Reynolds, J.D. 1999. Structural change in an exploited fish community: a consequence of differential fishing effects on species with contrasting life histories. Journal of Animal Ecology, 68, 617–627.

Jennings, S.; Kaiser, M.J. 1998. The effects of fishing on marine ecosystems. Advances in Marine Biology, 34, 201–352.

Jennings, S.; Reynolds, J.D.; Mills, S.C. 1997. Life history correlates of responses to fisheries exploitation. Proceedings of the Royal Society of London, 265, 333–339.

Jentoft, S. 1985. Models of fishery development: the cooperative approach. Marine Policy, 9, 322–331.

______ 1989. Fisheries co-management: delegating government responsibility to fishermen's organizations. Marine Policy, 13(2), 137–154.

______ 1999. Women in fisheries: beyond the veil. Samudra, 23, 3–6.

______ 2000. The community: a missing link of fisheries management. Marine Policy, 24, 53–59.

Jentoft, S.; Kristoffersen, T. 1989. Fishermen's co-management: the case of the Lofoten fishery. Human Organization, 48(4), 355–365.

Jentoft, S.; McCay, B.J. 1995. User participation in fisheries management. Lessons drawn from international experiences. Marine Policy, 19, 227–246.

Jentoft, S.; McCay, B.J.; Wilson, D.C. 1998. Social theory and fisheries co-management. Marine Policy, 22, 423–436.

Johannes, R.E. 1978. Traditional marine conservation methods in Oceania and their demise. Annual Reviews of Ecology and Systematics, 9, 349–364.

______ 1981. Words of the lagoon. Fishing and marine lore in the Palau District of Micronesia. University of California Press, Berkeley, CA, USA.

______ 1982. Traditional conservation methods and protected marine areas in Oceania. Ambio, 11, 258–261.

______ 1998a. The case for data-less marine resource management: examples from tropical nearshore fisheries. Trends in Ecology and Evolution, 13, 243–246.

______ 1998b. Government-supported, village-based management of marine resources in Vanuatu. Ocean & Coastal Management, 40, 165–186.

______ Lasserre, P.; Nixon, S.W.; Pliya, J.; Ruddle, K. 1983. Traditional knowledge and management of marine coastal systems. Biology International, Special issue 4.

Jorgensen, D. L. 1989. Participant observation. Sage Publications. London, UK.

Kapetsky, J.M. 1985. Some considerations for the management of coastal lagoon and estuarine fisheries. Food and Agriculture Organization, Rome, Italy. Fisheries Technical Paper No. 218.

Katon, B.; Pomeroy, R.S.; Salamanca, A. 1997. The marine conservation project for San Salvador: a case study of fisheries co-management in the Philippines, International Center for Living Aquatic Resources Management, Manila, Philippines. Fisheries Co-management Research Project Working Paper No. 23.

Katzer, J.; Cook, K.H.; Crouch, W.W. 1982. Evaluating information: a guide for users of social science research (2nd ed.). Addison-Wesley Publishing Company, Reading, MA, USA.

King, M. 2000. Fishers management by communities: a manual on promoting the management of subsistence fishers by Pacific Island communities. Secretariat of the Pacific Community, New Caledonia. 87 pp.

Kooiman, J.; Van Vliet, M.; Jentoft, S. 1999. Creative governance: opportunities for fisheries in Europe. Ashgate Publishing, Aldershot, UK.

Korten, D.C. 1986. Community-based resource management. *In* Korten, D.C., ed., Community management: Asian experiences and perspectives. Kumarian Press, West Hartford, CT, USA.

Kuperan, K. 1992. Deterrence and voluntary compliance with the regulation in the Malaysian fishery. University of Rhode Island, Kingston, RI, USA. PhD dissertation.

Kuperan, K.; Abdullah, N.M.R. 1994. Planning and management of small-scale coastal fisheries. *In* Pomeroy, R.S., ed., Community management and common property of coastal fisheries in Asia and the Pacific: concepts, methods and experiences. ICLARM (International Center for Living Aquatic Resources Management) Conference Proceedings 45. ICLARM, Manila, Philippines, pp. 115–123.

Kuperan, K.; Abdullah, N.; Susilowati, I.; Siason, I.; Ticao, C. 1997. Enforcement and compliance with fisheries regulations in Malaysia, Indonesia and the Philippines. ICLARM, Makati, Philippines. ICLARM Co-management Project Research Report No. 5,

Langill, S., compiler. 1999. Stakeholder analysis. Volume 7. Supplement for conflict and collaboration resource book. International Development Research Centre, Ottawa, ON, Canada.

Larkin, P.A. 1977. An epitaph for the concept of maximum sustained yield. Transactions of America. Fisheries Society, 106, 1–11.

Lees, S.H.; Bates, D.G. 1990. The ecology of cumulative change. *In* Moran, E.F., ed., The ecosystem approach in anthropology. University of Michigan Press, Ann Arbor, MI, USA. pp. 247–277.

Lieber, M.D. 1994. More than a living: fishing and the social order on a Polynesian atoll. Westview Press, Boulder, CO, USA.

Lim, C.P.; Matsuda, Y.; Shigemi, Y. 1995. Co-management in marine fisheries: the Japanese experience. Coastal Management, 23, 195–221.

Longhurst, A. 1998a. Ecological geography of the sea. Academic Press, New York, NY, USA.

Longhurst, A. 1998b. Cod: perhaps if we all stood back a bit? Fisheries Research, 38, 101–108.

Loucks, L.; Charles, A.T.; Butler, M., ed. 1998. Managing our fisheries, managing ourselves. Gorsebrook Research Institute, Halifax, NS, Canada.

Ludwig, D.; Hilborn, R.; Walters, C. 1993. Uncertainty, resource exploitation and conservation: lessons from history. Science, 260, 17-36.

Mace, P.M. 1994. Relationships between common biological reference points used as thresholds and targets of fisheries management strategies. Canadian Journal of Fisheries and Aquatic Sciences, 51, 110–122.

Mackinson, S. 2000. An adaptive fuzzy expert system for predicting structure, dynamics and distribution of herring shoals. Ecological Modelling, 126, 155–178.

Mackinson, S.; Nottestad, L. 1998. Combining local and scientific knowledge. Reviews in Fish Biology and Fisheries, 8, 481–490.

Magpayo, N. 1995. FIRMED (Fishery Integrated Resource Management for Economic Development): an integrated approach to community-based coastal resource management. *In* Foltz, C.; Pomeroy, R.S.; Barber, C.V., ed., Proceedings of the Visayas-wide conference on community-based coastal resources management and fisheries co-management. ICLARM (International Center for Living Aquatic Resources Management), Manila, Philippines. Fisheries Co-management Project Working Paper 4.

Mahon, R. 1986. Developing a management strategy for the flying-fish fishery of the eastern Caribbean. Proceedings of the Gulf and Caribbean Fisheries Institute, 39, 402.

Mahon, R., ed. 1987. Report and proceedings of the expert consultation on shared fishery resources of the Lesser Antilles region. Food and Agriculture Organization, Rome, Italy. FAO Fisheries Report No. 383.

______ 1990. Fishery management options for Lesser Antilles countries. FAO, Rome, Italy. FAO Fisheries Technical Paper No. 313.

______ 1991. Developing fishery data collection systems for eastern Caribbean islands. Proceedings of the Gulf Caribbean Fish Institute, 40, 309–322.

______ 1997. Does fisheries science serve the needs of managers of small stocks in developing countries? Canadian Journal of Fisheries and Aquatic Sciences, 54, 2207–2213.

______ 2001. Bigger brush, bolder strokes: management objectives and reference points for Caribbean coral reef fish fisheries. Coral Reefs (in preparation).

Maine, R.A.; Cam, B; Davis-Case, D. 1996. Participatory analysis, monitoring and evaluation for fishing communities: a manual. Food and Agricultural Organization, Rome, Italy. FAO Fisheries Technical Paper No. 364.

Margoluis, R.; Salafsky, N. 1998. Measures of success: designing, managing, and monitoring conservation and development projects. Island Press, Washington, DC, USA.

Matsuda, Y.; Kaneda, Y. 1984. The seven greatest fisheries incidents in Japan. *In* Ruddle, K; Akimichi, T., ed., Maritime institutions in the Western Pacific. Senri Ethnological Studies 17. National Museum of Ethnology, Osaka, Japan. pp. 159–182.

Matthews, D.R. 1988. Controlling common property: regulating Canada's east coast fishery. University of Toronto Press, Toronto, ON, Canada.

McCay, B.J.; Acheson, J.M., ed. 1987. The question of the commons. University of Arizona Press, Tucson, AZ, USA.

McCay, B.J.; Jentoft, S. 1996. From the bottom up: participatory issues in fisheries management. Society and Natural Resources, 9(3), 237–250.

______ 1998. Market or community failure? Critical perspectives on common property research. Human Organization, 57, 21–29.

McCay, B.J.; Wilson, D. 1998. How the participants talk about "participation" in mid-Atlantic fisheries management. Ocean and Coastal Management, 41, 41–69.

McConney, P.A. 1997. Social strategies for coping with uncertainty in the Barbados small-scale pelagic fishery. Proceedings of the Gulf and Caribbean Fisheries Institute, 49, 99–113.

______ 1998. Using "common science" in co-management. Proceedings of the Gulf and Caribbean Fisheries Institute, 50, 1115–1121.

McConney, P.A.; Atapattu, A.; Leslie, D. 1998. Organizing fisherfolk in Barbados. Proceedings of the Gulf and Caribbean Fisheries Institute, 51, 299-308.

McConney, P.A.; Mahon, R. 1998. Introducing fishery management planning to Barbados. Ocean and Coastal Management, 39, 189–195.

McEvoy, A.F. 1986. The fishermen's problem: ecology and law in the California Fisheries, 1850–1980. Cambridge University Press, Cambridge, UK.

McGoodwin, J.R. 1990. Crisis in world's fisheries: people, problems and politics. Stanford University Press, Stanford, CA, USA.

______ 1992. The case for co-operative co-management. Australian Fisheries, May 1992. pp. 11–15.

______ 1994. "Nowadays, nobody has any respect": the demise of folk management in a rural Mexican fishery. *In* Dyer, C.L.; McGoodwin, J.R., ed., Folk management in the world's fisheries. University Press of Colorado, Niwot, CO, USA. pp. 43–54.

McManus, J.W. 1997. Tropical marine fisheries and the future of coral reefs: a brief review with emphasis on Southeast Asia. Coral Reefs, 16, 121–127.

Meaden, G.J.; Do Chi, T. 1996. Geographical information systems: applications to marine fisheries. Food and Agriculture Organization, Rome, Italy. FAO Fisheries Technical Paper No. 356, 335 pp.

Miller, D. 1989. The evolution of Mexico's Caribbean spiny lobster fishery. *In* Berkes, F., ed., Common property resources: ecology and community-based sustainable development. Belhaven Press, London, UK.

Miller, R.J. 1990. Properties of a well-managed nearshore fishery. Fisheries, 15(5), 7–12.

Moffat, D.; Ngoile, M.N.; Linden, O.; Francis, J. 1998. The reality of the stomach: coastal management at the local level in Eastern Africa. Ambio, 27, 590–598.

Morrissey, M.T. 1989. Operational assessment of fisheries development projects. *In* Pollnac, R.B., ed., Monitoring and evaluating the impacts of small-scale fishery projects. International Center for Marine Resource Development, University of Rhode Island, Kingston, RI, USA. pp. 133–144.

Msiska, O.V.; Hersoug, S. 1997. Training in fisheries planning and management: the case of the Southern African Development Community nations. Naga, 20, 4–7.

Munro, G.; Bingham, N.; Pikitch, E. 1998. Individual transferable quotas, community-based fisheries management systems, and "virtual" communities. Fisheries, 23(3), 12–15.

Munro, J.L. 1979. Stock assessment models: applicability and utility in tropical small-scale fisheries. *In* Saila, S.B.; Roedel, P.M., ed., Stock assessment for tropical small-scale fisheries. International Center for Marine Resource Development, University of Rhode Island, Kingston, RI, USA. pp. 35–47.

Naroll, R. 1962. Data quality control — a new research technique. The Free Press of Glencoe, New York, NY, USA.

Navia, O.; Landivar, J. 1997. Resource book on participation. Inter-American Development Bank, Washington, DC, USA.

Neis, B. 1992. Fishers' ecological knowledge and stock assessment in Newfoundland. Newfoundland Studies, 8, 155–178.

Neis, B.; Felt, L.; Schneider, D.C.; Haedrich, R.; Hutchings, J.; Fischer, J. 1996. Northern cod stock assessment: what can be learned from interviewing resource users? Department of Fisheries and Oceans, Ottawa, ON, Canada. Atlantic Fisheries Research Document No. 96/45.

Newland, K. 1981. Infant mortality and the health of societies. Worldwatch Institute, Washington, DC, USA.

Nielsen, J.; Vedsmand, T. 1999. User participation and institutional change in fisheries management: a viable alternative to the failures of "top-down" driven control? Ocean and Coastal Management, 42 (1), 19-47.

Normann, A.K.; Raakjaer-Nielsen, J.; Sverdrup-Jensen, S., ed. 1998. Fisheries co-management in Africa: proceedings from a regional workshop on fisheries co-management research. Institute for Fisheries Management, North Sea Centre, Hirtshals, Denmark.

North, D.C. 1993. Economic performance through time. The Nobel Foundation and the Royal Swedish Academy of Sciences, Stockholm, Sweden. The Nobel Prizes 1993.

Novaczek, I.; Harkes, I. 1998. An institutional analysis of Sasi in Maluku, Indonesia. ICLARM (International Center for Living Aquatic Resources Management), Manila, Philippines. Fishery Co-management Research Project Working Paper No. 39.

NRC (National Research Council). 1999. Sustaining marine fisheries. Committee for Ecosystem Management for Sustainable Marine Fisheries, NRC. National Academy Press, Washington, DC, USA.

NRTEE (National Round Table on the Environment and the Economy). 1998. Sustainable strategies for oceans: a co-management guide. NRTEE, Ottawa, ON, Canada.

Ohtsuka, R.; Kuchikura, Y. 1984. The comparative ecology of subsistence and commercial fishing in southwestern Japan, with special reference to Maritime institutions. *In* Ruddle, K.; Akimichi, T. ed., Maritime institutions in the Western Pacific. Senri Ethnological Studies 17. National Museum of Ethnology, Osaka, Japan. pp. 121–136.

Ostrom, E. 1990. Governing the Commons: the evolution of institutions for collective action. Cambridge University Press, Cambridge, UK.

______ 1992. Crafting institutions for self-governing irrigation systems. Institute for Contemporary Studies, San Francisco, CA, USA.

______ 1994. Institutional analysis, design principles and threats to sustainable community governance and management of commons. *In* Pomeroy, R.S., ed., Community management and common property of coastal fisheries and Asia and the Pacific: concepts, methods and experiences. ICLARM (International Center for Living Aquatic Resources Management) Conference Proceedings 45. ICLARM, Manila, Philippines. pp. 34–50.

Ostrom, E.; Burger, J.; Field, C.B.; Norgaard, R.B.; Policansky, D. 1999. Revisiting the commons: local lessons, global challenges. Science, 284, 278–282.

Panayotou, T. 1982. Management concepts for small-scale fisheries: economic and social aspects. Food and Agriculture Organization, Rome, Italy. FAO Fisheries Technical Paper 228.

Pauly, D. 1979. A new methodology for rapidly acquiring basic information on tropical fish stocks: growth, mortality, and stock recruitment relationships. *In* Saila, S.B.; Roedel, P.M., ed., Stock assessment for tropical small-scale fisheries. International Center for Marine Resource Development, University of Rhode Island, RI, USA. pp. 154–172.

______ 1983. Some simple methods for the assessment of tropical fish stocks. Food and Agriculture Organization, Rome, Italy. FAO Fisheries Technical Paper No. 234.

______ 1994. On the sex of fish and the gender of scientists: a collection of essays in fisheries science. Chapman and Hall, London, UK.

Pelto, P.J.; Pelto, G.H. 1978. Anthropological research: the structure of inquiry. Cambridge University Press, Cambridge, UK.

Pido, M.D. 1995. Toward improved linkage between research and management in marine fisheries. Naga, 18, 15–18.

Pido, M.D.; Pomeroy, R.S.; Carlos, M.B.; Garces, L.R. 1996. A handbook for rapid appraisal of fisheries management systems. International Center for Living Aquatic Resources Management, Makati City, Philippines.

Pinkerton, E., ed. 1989a. Cooperative management of local fisheries: new directions for improved management and community development. University of British Columbia Press, Vancouver, BC, Canada.

______ 1989b. Introduction: attaining better fisheries management through co-management — prospects, problems, and propositions. *In* Pinkerton, E., ed., Cooperative management of local fisheries: new directions for improved management and community development. University of British Columbia Press, Vancouver, BC, Canada.

______ 1994. Summary and conclusions. *In* Dyer, C.L.; McGoodwin, J.R., ed., Folk management in the world's fisheries. University Press of Colorado, Niwot, CO, USA. pp. 317–337.

Pitcher, T.J. 1999. Rapfish, a rapid appraisal technique for fisheries, and its application to the code of conduct for responsible fisheries. Food and Agriculture Organization, Rome, Italy. Fish Circular No. 947.

______ 2000. The design and monitoring of marine reserves. University of British Columbia, Vancouver, BC, Canada. Fisheries Center Research Report No. 5.

Pitcher, T.J.; Bundy, A.; Preikshot, D.; Hutton, T.; Pauly, D. 1998. Measuring the unmeasurable: a multivariate and interdisciplinary method for rapid appraisal of the health of fisheries. *In* Pitcher, T.J.; Hart, P.J.B.; Pauly, D., ed., Reinventing Fisheries Management. Kluwer Academic Publishers, London, UK. pp. 31–54.

Pitcher, T.J.; Hart, P.J.B.; Pauly, D., ed. 1998. Reinventing fisheries management. Kluwer, London, UK.

Pitcher, T.J.; Pauly, D. 1998. Rebuilding ecosystems, not sustainability, as the proper goal of fishery management. *In* Pitcher, T.J.; Hart, P.J.B.; Pauly, D., ed., Reinventing fisheries management. Kluwer Academic Publishers, London, UK. pp. 311–329.

Pitcher, T.J.; Preikshot, D. 2000. Rapfish: a rapid appraisal technique to evaluate the sustainability status of fisheries. Fisheries Research, in press.

Poggie, J.J. 1972. Toward quality control in key informant data. Human Organization, 31(1), 23–30.

Poggie, J.J.; Pollnac, R.B., ed. 1991a. Small-scale fisheries development: sociocultural perspectives. International Center for Marine Resource Development, University of Rhode Island, Kingston, RI, USA.

______ 1991b. Community context and cooperative success in Ecuador. *In* Poggie, J.J.; Pollnac, R.B., ed., Small-scale fishery development: sociocultural perspectives. International Center for Marine Resource Development, University of Rhode Island, Kingston, RI, USA. pp. 106–113.

Pollnac, R.B. 1974. Intracultural variability in the conceptualization of food plants in Buganda. Makerere Institute of Social Research, Makerere University, Kampala, Nkanga, Uganda 1974.

______ 1980. The ethnoicthyology of small-scale fishermen of Puntarenas, Costa Rica. *In* Sutinen, J.G.; Pollnac, R.B., ed., Small-scale fisheries in Central America: acquiring information for decision-making. International Center for Marine Resource Development, Kingston, RI, USA.

______ 1984. Investigating territorial use rights among fishermen. *In* Ruddle, K.; Akimichi, T., ed., Maritime Institutions in the Western Pacific. Senri Ethnological Studies 17, National Museum of Ethnology, Osaka, Japan. pp. 285–300.

______ 1988. Evaluating the potential of fishermen's organizations in developing countries. International Center for Marine Resource Development, University of Rhode Island, Kingston, RI, USA.

______ 1994. Research directed at developing local organizations for people's participation in fisheries management. *In* Pomeroy, R.S., ed., Community management and common property of coastal fisheries in Asia and the Pacific: concepts, methods and experiences. International Center for Living Aquatic Resources Management, Manila, Philippines. pp. 94–106.

______ 1998. Rapid assessment of management parameters for coral reefs. Coastal Management Report No. 2205 and ICLARM (International Center for Living Aquatic Resources Management) Contribution No. 1445. Coastal Resources Center, University of Rhode Island and ICLARM, Narragansett, RI, USA.

Pollnac, R.B.; Dickson, A.; Sualog, A.; Razo, N.; RFTC Project Coordinating Staff. 1989. Evaluation of two small-scale fishermen's training projects in Palawan. *In* Pollnac, R.B., ed., Monitoring and evaluating the impacts of small-scale fishery projects. International Center for Marine Resource Development, University of Rhode Island, Kingston, RI, USA. pp. 133–144.

Pollnac, R.B.; Gorospe, M.L.G. 1998. Aspects of the human ecology of Atulayan Bay. *In* Pollnac, R.B., ed., Rapid assessment of management parameters for coral reefs. Coastal Management Report No. 2205 and ICLARM (International Center for Living Aquatic Resources Management) Contribution No.1445. Coastal Resources Center, University of RI and ICLARM, Narragansett, RI, USA. pp. 77–139.

Pollnac, R.B.; Poggie, J.J. 1988. The structure of job satisfaction among New England fishermen and its application to fisheries management policy. American Anthropologist, 90, 888–901.

______ 1997. Fish aggregating devices in developing countries: problems and perspectives. International Center for Marine Resource Development and Anthropology Program, University of Rhode Island, Kingston, RI, USA.

Pollnac, R.B.; Pomeroy, R.S. 1996. Evaluating factors contributing to the success of community-based coastal resource management projects: a baseline independent method. Department of Anthropology and Coastal Resources Center, University of Rhode Island, Kingston, RI, USA. Anthropology Working Paper No. 54.

Pollnac, R.B.; Sihombing, S. 1996. Cages, controversy and conflict: carp culture in Lake Toba, Indonesia. *In* Bailey, C.; Jentoft, S.; Sinclair, P., ed., Aquaculture development: social dimensions of an emerging industry. Westview Press, Boulder, CO, USA.

Pollnac, R.B.; Sondita, F.; Crawford, B.; Mantjoro, E.; Rotinsulu, C.; Siahainenia, A. 1997. Socioeconomic aspects of resource use in Bentenan and Tumbak. Coastal Resources Center, University of Rhode Island, Kingston, RI, USA.

Pomeroy, R.S. 1989. Monitoring and evaluation of fishery and agriculture projects: case study and discussion of issues. *In* Pollnac, R.B., ed., Monitoring and evaluating the impacts of small-scale fishery projects. International Center for Marine Resource Development, University of Rhode Island, Kingston, RI, USA. pp 41–55.

______ 1994a. Introduction. *In* Pomeroy, R.S., ed., Community management and common property of coastal fisheries in Asia and the Pacific: concepts, methods and experiences. ICLARM (International Center for Living Aquatic Resources Management) Conference Proceedings 45. ICLARM, Manila, Philippines. pp. 1–11.

______ ed. 1994b. Community management and common property of coastal fisheries in Asia and the Pacific: concepts, methods and experiences. ICLARM (International Center for Living Aquatic Resources Management), Manila, Philippines.

______ 1995. Community-based and co-management institutions for sustainable coastal fisheries management in Southeast Asia. Ocean and Coastal Management, 27(3), 143–162.

Pomeroy, R.S.; Berkes, F. 1997. Two to tango: the role of government in fisheries co-management. Marine Policy, 21, 465–480.

Pomeroy, R.S.; Carlos, M.B. 1997. Community-based coastal resource management in the Philippines: a review and evaluation of programs and projects, 1984–1994. Marine Policy, 21, 445–464.

Pomeroy, R.S.; Katon, B.; Genio, E.; Harkes, I. 1999. Fisheries co-management in Asia: lessons from experience. ICLARM (International Center for Living Aquatic Resources Management), Manila, Philippines.

Pomeroy, R.S.; Pido, M. 1995. Initiatives toward fisheries co-management in the Philippines: the case of San Miguel Bay. Marine Policy, 19, 213–226.

Pomeroy, R.S.; Pollnac, R.B.; Katon, B.M.; Predo, C.D. 1997. Evaluating factors contributing to the success of community-based coastal resource management: the Central Visayas regional Project-1, Philippines. Ocean and Coastal Management, 36(1-3), 97–120.

Pomeroy, R.S.; Williams, M.J. 1994. Fisheries co-management and small-scale fisheries: a policy brief. ICLARM (International Center for Living Aquatic Resources Management), Manila, Philippines.

Prado, J. 1997. Responsible fisheries with specific reference to small-scale fisheries and West Africa. Fisheries Department, Food and Agriculture Organization, Rome, Italy.

Renard, Y. 1991. Institutional challenges for community-based management in the Caribbean. Nature and Resources, 27(4), 4–9.

Ricci, J.A.; Jerome, N.W.; Megally, N.; Galal, O.; Harrison, G.G.; Kirksey, A. 1995. Assessing the validity of informant recall: results of a time use pilot study in peri-urban Egypt. Human Organization, 54(3), 304–308.

Rivera, R.A. 1997. Re-inventing power and politics in coastal communities: community-based and coastal resource management in the Philippines. Marine Affairs Program, Dalhousie University, Halifax, NS, Canada.

Rivera, R.A.; Newkirk G.F. 1997. Power from the people: a documentation of non-governmental organizations' experience in community-based coastal resource management in the Philippines. Ocean and Coastal Management, 36, 73–95.

Roberts, C.M.; Hawkins, J.P. 1999. Extinction risk in the sea. Trends in Ecology and Evolution, 14, 241–246.

Robinson, J.; Athanisou, R.; Head, K. 1969. Measures of occupational attitudes and occupational characteristics. Institute for Social Research, University of Michigan, Ann Arbor, MI, USA.

Roedel, P.M., ed. 1975. Optimum sustainable yield as a concept in fisheries management. American Fisheries Society, Bethesda, MD, USA. Special Publication No. 9.

Rosander, A.C. 1977. Case studies in sample design. M. Dekker, New York, NY, USA.

Rosenberg, A.A.; Restrepo, V.R. 1994. Uncertainty and risk evaluation in stock assessment advice for U.S. marine fisheries. Canadian Journal of Fisheries and Aquatic Sciences, 51, 2715–2720.

Ruddle, K. 1987. Administration and conflict management in Japanese coastal fisheries. Food and Agriculture Organization, Rome, Italy. FAO Fisheries Technical Paper No. 273.

______ 1988. Social principles underlying traditional inshore fishery management systems in the Pacific Basin. Marine Resource Economics, 5, 351–363.

______ 1989. Solving the common-property dilemma: village fisheries rights in Japanese coastal waters. *In* Berkes, F., ed., Common property resources, Belhaven Press, London, UK. pp. 168–184.

______ 1993. External forces and change in traditional community-based fishery management systems in the Asia-Pacific region. MAST, 6, 1–37.

______ 1994. Changing the focus of coastal fisheries management. *In* Pomeroy, R.S., ed., Community management and common property of coastal fisheries in Asia and the Pacific: concepts, methods and experiences. ICLARM (International Center for Living Aquatic Resources) Conference Proceedings 45. ICLARM, Manila, Philippines.

Ruddle, K.; Akimichi, T., ed. 1984. Maritime institutions in the Western Pacific. National Museum of Ethnology, Senri Ethnological Studies 17, Osaka, Japan.

Ruddle, K.; Hviding, E.; Johannes, R.E. 1992. Marine resource management in the context of customary tenure. Marine Resource Economics, 7(4), 249–273.

Sachs, J. 1999. Sachs on development: helping the world's poorest. The Economist, 14 Aug. 1999, pp. 17–20.

Salm, R.V.; Clark, J.R. 2000. Marine and coastal protected areas. A guide for planners and managers (2nd ed.). International Conservation Union, Gland, Switzerland.

Sary, Z.; Oxenford, H.A.; Woodley, J.D. 1997. Effects of an increase in trap mesh size on an overexploited coral reef fishery at Discovery Bay, Jamaica. Marine Ecology Progress Series, 154, 107–120.

Schensul, J.J.; LeCompte, M.D. 1999. Ethnographer's toolkit (7 volumes). Altamira Press, Walnut Creek, CA, USA.

Schwartz, N.B. 1986. Socioeconomic considerations. *In* Lovshin, et al., ed., Cooperatively managed Panamanian rural fish ponds. International Center for Aquaculture Research and Development Series No. 33. Auburn University, Auburn, Alabama, USA.

Scott, A.D. 1955. The fishery: the objectives of sole ownership. Journal of Political Economy, 63, 116–124.

SEACAM (Secretariat for Eastern African Coastal Area Management). 1999. From a good idea to a successful project. A manual for development and management of local level projects. SEACAM, Maputo, Mozambique.

Seixas, C.S. 2000. The Ibiriquera lagoon, Brazil: a resilient social-ecological system? Papers of the International Association for the Study of Common Property, June 2000, Bloomington, Indiana, USA. http://www.indiana.edu/~iascp2000.htm.

Sen, S.; Raakjaer-Nielsen, J. 1996. Fisheries co-management: a comparative analysis. Marine Policy, 20, 405–418.

Shea, K. 1998. Perspectives: management of populations in conservation, harvesting and control. Trends in Ecology and Evolution, 13, 371–375.

Shepard, M.P. 1991. Fisheries research needs of small island countries. International Centre for Ocean Development, Halifax, NS, Canada.

Sherman K. 1992. Monitoring and assessment of large marine ecosystems: a global and regional perspective (chapter 59). *In* McKenzie, D.H.; Hyatt, D.E.; McDonald, V.J., ed., Ecological indicators, Vol. 1 and 2. Elsevier Press, Essex, England. pp. 1041–1074.

Sherman, K.; Alexander, L.M.; Gold, B.D. 1993. Large marine ecosystems, stress, mitigation and sustainability. AAAS Press, Washington, DC, USA.

Smith, A.H.; Berkes, F. 1991. Solutions to the "tragedy of the commons": sea-urchin management in St. Lucia, West Indies. Environmental Conservation, 18, 131–136.

______ 1993. Community-based use of mangrove resources in St. Lucia. International Journal of Environmental Studies, 43, 123–131.

Smith, I.R. 1979. A research framework for traditional fisheries. International Center for Living Aquatic Resources Management, Manila, Philippines. Studies and Reviews No. 2.

Smith, S.J. 1993. Risk evaluation and biological reference points for fisheries management: a review. *In* Kruse, G.; Eggers, D.M.; Marasco, R.J.; Pautzke, C.; Quinn II, T.J., ed., Management strategies for exploited fish populations. Alaska Sea Grant College Program, University of Alaska, Fairbanks, AK, USA. pp. 339–353.

Sparre P.; Venema, S.C. 1992a. Introduction to tropical fish stock assessment. Part 1 – manual. Food and Agriculture Organization, Rome, Italy. FAO Fisheries Technical Paper No. 306/1.

______ 1992b. Introduction to tropical fish stock assessment. Part 2 – manual. Food and Agriculture Organization, Rome, Italy. FAO Fisheries Technical Paper No. 306/2.

Spencer, L.J. 1989. Winning through participation. Kendall/Hunt, Iames, IA, USA.

Spradley, J. 1969. The ethnographic interview. Holt, Rinehart, and Winston, New York, NY, USA.

Steele, J.H. 1998. Regime shifts in marine ecosystems. Ecological Applications, 8(1), S33–S36.

Stevenson, D.; Pollnac, R.B.; Logan, P. 1982. A guide for the small-scale fishery administrator: information from the harvesting sector. International Center for Marine Resource Development, University of Rhode Island, RI, USA.

Susilowati, I. 1998. Economics of regulatory compliance in the fisheries of Indonesia, Malaysia and the Philippines. University of Putra, Malaysia. Unpublished thesis.

Sutinen, J.; Anderson, P. 1985. The economics of fisheries law enforcement. Land Economics, 61, 387–97.

Sutinen, J.; Kuperan, K. 1994. A socioeconomic theory of regulatory compliance in fisheries. Paper presented at the 7th conference of the International Institute of Fisheries Economics and Trade, July 1994, Taipei, Taiwan. International Institute of Fisheries Economics and Trade, Taipei, Taiwan.

Swezey, S.L.; Heizer, R.F. 1977. Ritual management of salmonid fish resources in California. Journal of California Anthropology, 4, 6–29.

Tam, J.; Palma, W.; Riofrio, M.; Aracena, O.; Lépez, M.I. 1996. Decision analysis applied to the fishery of the sea snail *Concholepas concholepas* from the central northern coast of Chile. Naga, 19, 45–48.

Tripartite Partnership in Marine and Aquatic Resources Management and Rural Development Program. n.d. Philippine Partnership for the Development of Human Resources for Rural Areas. Quezon City, Philippines.

Turner, M.G.; Dale, V.H. 1998. Comparing large, infrequent disturbances: what have we learned? Ecosystems, 1, 493–496.

Tyler, T. R. 1990. Why people obey the law. Yale University Press, New Haven, CT, USA.

United Nations. 1983. The law of the sea. Official text of the United Nations Convention on the Law of the Sea with annexes and tables. United Nations, New York, NY, USA.

______ 1992. Agenda 21: program of action for sustainable development. Final text of agreements negotiated by governments at the UNCED (United Nations Conference on Environment and Development), 3–14 June 1992, Rio de Janeiro, Brazil. United Nations, New York, NY, USA.

______ 1995. Agreement for the implementation of the provisions of the United Nations Convention of the Law of the Sea, 10 Dec. 1982, relating to the conservation and management of straddling fish stocks and highly migratory fish stocks. UN Conference on Straddling Fish stocks and Highly Migratory Species, Sixth session, A/Conf.164/37, United Nations, New York, NY, USA.

Uphoff, N. 1991. Fitting projects to people. *In* Cernea, M.M., ed., Putting people first: sociological variables in rural development (2nd ed.). Oxford University Press, New York, NY, USA. pp. 467–511.

USAID (US Agency for International Development). 1994. The logical framework: a project-level design tool. Document from a videotape presentation produced by the Professional Studies and Career Development Division, USAID and PASITAM (Program of Advanced Studies in Institution Building and Technical Assistance Methodology) of the MUCIA (Midwest Universities Consortium for International Activities), Washington, DC, USA.

Van Mulekom, L. 1999. An institutional development process in community-based coastal resource management. Ocean and Coastal Management, 42, 439–456.

Vestergaard, T.A. 1991. Living with pound nets: diffusion, invention and implications of a technology. Folk, 33, 149–167.

Walters, C.J. 1986. Adaptive management of renewable resources. McGraw-Hill, New York, NY, USA.

______ 1998. Designing fisheries management systems that do not depend on stock assessment. *In* Pitcher, T.J.; Hart, P.J.B.; Pauly, D., ed., Reinventing fisheries management. Kluwer Academic Publishers, London, UK. pp. 279–288.

Walters, J.S.; Maragos, J.; Siar, S.; White, A.T. 1998. Participatory coastal resource assessment: a handbook for community workers and coastal resource managers. Coastal Resource Management Project and Silliman University, Cebu City, Philippines.

Weinstein, M.S. 2000. Pieces of the puzzle: solutions for community-based fisheries management from native Canadians, Japanese cooperatives and common property researchers. Georgetown International Environmental Law Review, 12(2), 375–412.

Weitzner, V.; Fonseca Borras, M. 1999. Cahuita, Limon, Costa Rica: from conflict to collaboration. *In* Buckles, D., ed., Cultivating peace: conflict and collaboration in natural resource management. IDRC (International Development Research Centre), Ottawa, ON, Canada. pp. 129–150.

Welcomme, R.L. 1998. Evaluation of stocking and introductions as management tools. *In* Cowx, I.G., ed., Stocking and introduction of fish. Fishing News Books, London, UK. pp. 397–413.

______ 1999. A review of a model for qualitative evaluation of exploitation levels in multi-species fisheries. Fisheries Management and Ecology, 6, 1–19.

White, A.T.; Hale, L.Z.; Renard, Y.; Cortes, L. 1994a. The need for community-based coral reef management. *In* White, A.T.; Hale, L.Z.; Renard, Y.; Cortes, L., ed., Collaborative and community-based management of coral reefs. Kumarian Press, West Hartford, CN, USA. pp. 1–18.

______ ed. 1994b. Collaborative and community-based management of coral reefs. Kumarian Press, West Hartford, CT, USA.

Williams, M. 1996. The transition in the contribution of living aquatic resources to food security. International Food Policy Research Institute, Washington, DC, USA. Food, Agriculture and the Environment Discussion Paper 13.

Wilson, J.A.; Acheson, J.M.; Metcalfe, M.; Kleban, P. 1994. Chaos, complexity and community management of fisheries. Marine Policy, 184, 291–305.

World Bank. 1999. Voices from the village: a comparative study of coastal resource management in the Pacific Islands. The World Bank, Washington, DC, USA. Draft report (not for circulation). Pacific Islands Discussion Paper Series, No. 9, East Asia and Pacific Region, Papua New Guinea and Pacific Islands Country Management Unit.

World Bank, United Nations Development Programme, Commission of Environmental Corporation, and Food and Agriculture Organization. 1992. A study of international fisheries research. The World Bank, Washington, DC, USA. 103 pp.

Zerner, C. 1994. Tracking sasi: the transformation of a central Molucan reef management institution in Indonesia. *In* White, A.T.; Hale, L.Z.; Renard, Y.; Cortes, L., ed. Collaborative and community-based management of coral reefs. Kumarian Press, West Hartford, CN, USA. pp. 19–32.

Index

C

F

G

H

I

J

K

L

M

N

O

P

Q

R

S

T

U

V

W

Y

About the Institution

The International Development Research Centre (IDRC) is committed to building a sustainable and equitable world. IDRC funds developing-world researchers, thus enabling the people of the South to find their own solutions to their own problems. IDRC also maintains information networks and forges linkages that allow Canadians and their developing-world partners to benefit equally from a global sharing of knowledge. Through its actions, IDRC is helping others to help themselves.

About the Publisher

IDRC Books publishes research results and scholarly studies on global and regional issues related to sustainable and equitable development. As a specialist in development literature, IDRC Books contributes to the body of knowledge on these issues to further the cause of global understanding and equity. IDRC publications are sold through its head office in Ottawa, Canada, as well as by IDRC's agents and distributors around the world. The full catalogue is available at http://www.idrc.ca/booktique/.